The Ontario Economy 1982-1995

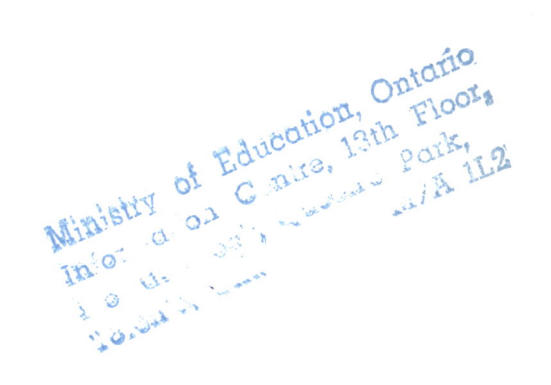

Policy Study Series

Ontario Economic Council

Peter Dungan
Douglas Crocker
Gay M. Garesché

Volume 2

© Ontario Economic Council 1983
81 Wellesley Street East
Toronto, Ontario
M4Y 1H6

Printed in Canada

ISBN Complete 2 volume set: 7586-8
ISBN Volume II: 8096-9

ISSN 0227-0005

Peter Dungan is an assistant professor of economics at the Institute for Policy Analysis, University of Toronto, where he is also the models manager of the Policy and Economic Analysis Program.

Douglas Crocker is a research officer at the Ontario Economic Council.

Gay M. Garesché is an economist at the University of Toronto.

This report reflects the views of the authors and not necessarily those of the Ontario Economic Council or the Government of Ontario. The Council establishes policy questions to be investigated and commissions research projects, but it does not influence the conclusions or recommendations of authors. The decision to sponsor publication of this study was based on its competence and relevance to public policy.

Contents
Volume 2

Introduction   1

2
The national outlook - FOCUS   3

3
National industrial detail - PRISM   13

4
Provincial projection - PRISM   17

5
Provincial projection: overview of results   35

6
Results by province   46

7
Appendix - tables   60

List of Tables   157

List of Figures   159

1
Introduction

Since 1976 the Institute for Policy Analysis, through the Policy and Economic Analysis Program, has become increasingly involved in modelling the provincial economies and in regional policy analysis. In early 1981 PEAP introduced PRISM (the Provincial Industrial Satellite Model) which, in tandem with the FOCUS national macroeconometric model, is intended for use in regional policy analysis and longer-term projection.

PRISM has now been used to generate several projections for the provinces - most notably for the policy study commissioned by the Ontario Economic Council (OEC). Although the OEC study focuses on Ontario results, it does provide a viable scenario for Canada and the provinces through 1990. While events have overtaken even this recent projection,[*] it does offer one possible, albeit relatively optimistic, outlook for the Canadian economic environment.

This Policy Study presents the results of a more recent PRISM projection initially discussed at the PEAP Conference on the Regional Economies held 10 September 1982. It is intended to provide an up-to-date outlook for the national and provincial economies incorporating current data and the present policy environment.

For this outlook FOCUS was tuned initially to the Data Resources (DRI) low-trend long-term solution of February 1982, which was based on the confluence of relatively pessimistic inflation, growth and policy assumptions and trends. While the FOCUS results evolved somewhat differently than the DRI

---

[*] This study was concluded in the spring of 1982, but has been somewhat delayed in publication.

The Ontario Economy 1982 - 1995

results, the inclusion of the poor economic developments established by the second quarter national accounts and the August Labour Force Survey added to the pessimism of this scenario. In addition, this outlook includes a number of somewhat gloomy assumptions intended to produce a 'low-trend' case. These assumptions will be noted throughout the text.

What follows is a brief description of the September outlook's national assumptions and results to 1995. This is followed by a more complete exposition of the long-term industry and provincial assumptions and results. Detailed tables of the results are provided in the appendix.

2
The national outlook — FOCUS

ASSUMPTIONS

International Environment: Most important in the international environment are developments in the U.S. (see Table 1). While the point of departure for the FOCUS outlook was the long-term low-trend spring DRI forecast for the U.S., selective adjustments in subsequent FOCUS runs have caused it to diverge somewhat from the DRI assumptions. The principal features of the U.S. outlook which emerge are as follows:

- The U.S. economy experiences sluggish growth in total real output in the next two years followed by a slow recovery. Growth remains well below potential through 1983 and at or only slightly above potential throughout the recovery period.
- The U.S. fails to improve markedly upon the inflation experience of the 1970s and early 1980s. This is a key pessimistic assumption of this scenario.
- The Federal Reserve reinstates relatively severe monetary restraint during the winter of 1983 to avoid backpedalling on inflation, undercutting real growth.

Other assumptions concerning the external environment are:

- Canada's other major trade partners suffer similar economic malaise.
- The Canadian dollar remains stable at about 81¢ U.S. as a good balance

The Ontario Economy 1982 - 1995

TABLE 1
National outlook assumptions

	1962 -73	1974 -80	1981 -85	1986 -90	1991 -95
U.S. real GNE average annual growth	4.4	2.6	1.8	3.0	1.9
U.S. GNP deflator average annual growth	3.7	7.7	8.2	10.2	9.7
Population average annual growth	1.6	1.2	.9	.8	.6
Population 15+ average annual growth	2.4	2.1	1.4	1.0	.9

of trade performance in 1982, a stable gap between Canadian and U.S. interest rates in the next few years and a favourable energy balance in the late 1980s and early 1990s offset the effect of higher Canadian inflation and the pursuit of somewhat nationalistic economic policies.

. No major international financial shock is assumed.

Energy: The energy sector continues to function as per the NEP and 1981 Pricing Accords. Table 2 lists major energy projects included in the scenario.

. Real oil prices increase 3 to 5 per cent per annum after 1985 vis-à-vis the U.S. GNP deflator. (Zero real increase in the near term.)
. Poor energy product demand reflects the slow economy and conservation efforts. Slow energy supply growth reflects substantial cutbacks in investment in the early years of the scenario. Energy self-sufficiency is achieved by 1995-96.

Population: Population growth (see Table 1) slows throughout the projection horizon due to reduced immigration and a lower natural increase.

. Government targets for immigration are lowered in reaction to high

TABLE 2
Provinces - major projects

		On stream
Newfoundland	Hibernia	87-88
Nova Scotia	Sable Island gas	84-86
Saskatchewan	Tertiary recovery and gas production	early and mid 90s
Alberta	One new SYNCRUDE-type tarsands plant	early 90s
	'New' conventional crude and gas development	all years
British Columbia	Gas and port development	84-86
	Additional gas development	early 90s
Territories	Beaufort oil production, preliminary	95

domestic unemployment rates.

- The combination of low fertility rates and fewer individuals entering the prime child-bearing years leads to lower natural population increases.
- Growth in the population 15 years and older also decelerates, though not as fast as total population growth.
- High unemployment in the 1980s cuts into the large increases in the participation rate seen in the 1970s. This is due to a sizeable 'discouraged worker' effect.

Government Policy: The policy environment is generally assumed to continue as it now exists. However, this is also an area in which some relatively pessimistic assumptions have been made. For example, it has been assumed that there will be little significant spillover of the federal 6 and 5 per cent wage restraint program to the private sector. Also:

- Continued public sector wage and spending restraint is assumed beyond 1983, though it is not expected to be very vigorous.
- Effective personal tax increases are assumed by means of a cap on indexation of the personal tax structure.

The Ontario Economy 1982 - 1995

- UIC payments increase automatically to offset the large volume of payments made by the government to persons during the recession.
- The federal government reduces some income transfers to the provinces in an effort to reduce its deficit. On balance, however, the provinces remain less in deficit than the federal government.
- Monetary policy remains tight - but not so tight as to stifle a modest recovery despite assumed ongoing inflation.

On a brighter note:

- It is assumed that corporate tax receipts rise as corporate profits recover from extremely low recent levels.

The upshot of these public sector assumptions is that the federal government, in particular, manages to reduce its annual deficit markedly over the next decade and one half. Provincial governments also improve their financial position.

Inflation: Inflation continues virtually unabated at almost 9 per cent right through the early 1990s in this outlook. While inflation is obviously not an exogenous variable, it is influenced by many factors which are external to FOCUS. For example:

- Energy prices are assumed to escalate in real terms after 1985.
- Canadian wage determination is partially a reflection of social mores. There is only limited latitude for salary givebacks in this country like those which have occurred in U.S. heavy industry. Moreover, it is unclear, to some U.S. observers at least, that the much-touted U.S. 'givebacks' and wage reductions represent a fundamental change in U.S. labour practices. Instead, they argue that present wage and price conditions represent merely a typical reaction to a severe recession. (For example, see Daniel J.E. Mitchell, 'Recent Union Contract Concessions', Brookings Papers on Economics Activity (1/1982), pp. 165-201 and John T. Dunlop, 'Working Toward Consensus', Challenge July/August 1982, pp. 26-34. For an opposing view see Audrey Freeman, 'A

The National Outlook - FOCUS

Fundamental Change in Wage Bargaining', Challenge July/August 1982, pp. 14-17.)

- The large volume of Canadian output and consumption which is traded abroad makes this country particularly vulnerable to foreign inflation, which is assumed to be high in the U.S. and world economies in this scenario.

- The large number of non-market determined prices in Canada takes some inflation outside the realm of purely market solutions.

- It has also been assumed that present partly-voluntary, partly-mandatory restraints will have relatively little impact on wages and prices in the long run. (This assumption looks increasingly unlikely for the short run.)

An attempt has been made in this scenario to represent a number of the exogenous inflation-producing influences mentioned above. The result is high inflation which persists throughout the projection period despite the inflation fighting efforts of the government.

Several important results follow from these inflation assumptions, specifically, weak real growth, poor productivity performance and high unemployment. However, the intractable inflation problem is assumed to eventually force the government to change policy focus slightly from inflation to jobs. Only after this policy change does the recovery develop any momentum. The following provides more specific details of the results.

Results

Major results are summarized in Chart 1 and Tables 3 and 4. See also Appendix Tables 1-9.

- Real GNE recovers steadily though unspectacularly from the severe recession in 1982. Only in one year, 1985, is growth expected to exceed 5 per cent. The gap between actual and potential output opened during the recession is never closed in this scenario. Table 4 and Chart 2 provide a breakdown of GNE growth by demand category.

  . Consumption recovers vigorously following two poor years, 1982 and 1983. Growth subsides in the 1990s in tandem with population growth.

The Ontario Economy 1982 - 1995

TABLE 3
National projection - summary
Revised low-trend (per cent)

	1962 -73	1974 -80	1981 -85	1986 -90	1991 -95
Real GNP growth	5.8	2.7	1.9	3.3	2.6
Inflation (GNP deflator)	3.9	10.0	10.1	9.1	8.9
90-day paper rate	5.8	10.0	16.6	14.5	13.1
Ind. bond rate	7.3	10.8	16.3	14.8	14.3
Unemployment rate	4.9	7.3	10.3	9.0	7.4
Exchange rate (US$/Cdn.$)	.95	.93	.81	.81	.81

. Government spending remains steadfastly under 2 per cent growth per annum to 1995 as governments are continually constrained by their deficits.

. Investment spending after recovering from the 82-83 recession is the major 'engine' of growth. While not at levels foreseen several years ago, there is considerable investment in resource extraction and in upgrading existing stock. Except for two strong years in 1984 and 1985, housing revives slowly from the pre-1971 levels of activity experienced in 1982 and 1983. Non-residential construction and machinery and equipment investment benefit from major energy development activity in the mid-1980s and again in the early 1990s.

. The balance of trade on goods and services improves dramatically in 1982. Thereafter, the balance erodes slowly as Canada's inflation outpaces that of the U.S. while the Canadian dollar is held up by some resurgence of capital inflows.

- Productivity growth remains sluggish by historical standards. While rising real energy prices are felt by some to have had an important role in initiating the poor productivity performance of the last few years, more stable real energy prices are not assumed to have the reciprocal effect of bolstering productivity performance in this scenario. High inflation (especially through its impact via the tax system) has been suggested as another source of poor

Chart 1

Canada: Main Indicators

The Ontario Economy 1982 - 1995

TABLE 4
Components of real GNP - growth rates

	1962 -73	1974 -80	1981 -85	1986 -90	1991 -95
GNP	5.8	2.7	1.9	3.3	2.6
Consumption	5.5	3.6	1.8	3.6	2.7
Autos	9.7	1.7	-2.7	9.1	3.2
Durables excl. autos	9.5	4.8	1.6	4.2	3.7
Investment	6.9	3.5	0.4	3.8	4.7
Residential	7.7	-2.3	3.3	1.1	0.9
Non-res. struct.	4.6	6.6	1.6	4.5	5.9
Mach. & equip.	9.1	4.8	-2.1	4.5	5.0
Government	5.2	1.6	1.2	1.4	1.8
Exports	9.0	3.1	2.3	4.0	3.6
Imports	8.5	3.7	1.6	3.9	4.4

productivity performance. From this viewpoint, the persistence of inflation in this scenario is consistent with continued poor productivity.

Furthermore, the model is unable to distinguish that it is the 'weak sisters' or poor corporate performers which presumably have been weeded out by the current recession, thereby leaving the corporate sector stronger and more efficient. On the other hand, the recession may be so severe that even the strong sectors will be unable to generate major productivity growth. In addition, rather than letting the weak sectors die, government may have to support them because of their tendency to be large employers. The net effect is that productivity is projected to grow quite slowly.

- The unemployment rate remains above 11 per cent through 1984 and above 7 per cent until 1995. A gap between the actual and natural unemployment rate persists throughout the scenario. (The fact that inflation diminishes little in the face of these high unemployment rates suggests that indeed the inflation assumptions are pessimistic.)

- Interest rates remain essentially in the mid-teens, moderating only slightly in the early 1990s. Persistent inflation of about 9 per cent and the

Chart 2

Expenditure Shares of GNP

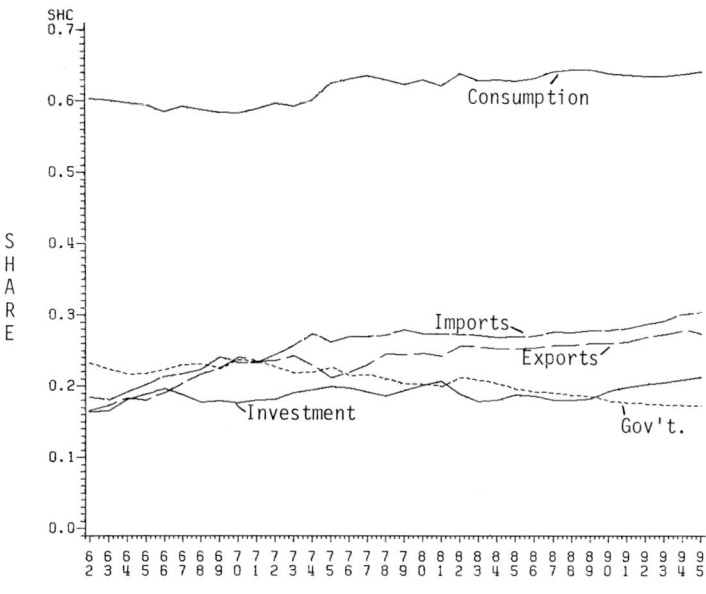

The Ontario Economy 1982 - 1995

maintenance of historically high real interest rates keep nominal rates high.

The composite picture is one of conflicting economic developments - high inflation and severe recession, poor productivity and high unemployment - with a single-minded policy intended to reduce inflation in the near term. Following a refocusing of government policy more on job formation, but without abandonment of the inflation fight, some economic progress is made later in the scenario.

Other policy environments can of course be envisaged. However, barring a government policy of 'growth at any cost' the real results may not be very different from those shown, assuming that inflation-fighting must remain a major focus of policy. Even assuming less exogenous inflationary pressure than included in this scenario, there is little latitude for expansionist fiscal and monetary policy and thus for substantially stronger real growth. Put otherwise, more exogenous 'room to manoeuvre' on inflation is most likely to be used to lower inflation and nominal (but not real) interest rates, and much less for additional real growth. So while this scenario includes pessimistic inflation assumptions, the end results are similar enough for the real variables to those for a less inflationary environment to make this scenario far more likely than a strong growth scenario.

3
National industrial detail – PRISM

The foregoing FOCUS results provide many of the key inputs to the industrial/provincial results developed with PRISM. In the first PRISM step the final demands calculated in FOCUS are converted by means of an input/output table and trend adjustment equations to national industry detail. Detailed industry output results are summarized in Table 5 and shown year by year in Appendix Tables 12-14. Shares of major industry groups in total output are shown in Chart 3.

Not surprisingly the goods-producing sectors suffer most from the current severe recession. Manufacturing output in particular takes until 1985 to recover its 1981 level (perhaps an overly pessimistic outcome), while its growth averages 3/4 of 1 per cent in the early 1990s. Service output, on the other hand, grows quite steadily for the next decade and one half except for the 4 plus per cent decline in 1982.

- Output in the primary sector falters seriously in the face of slack primary goods demand. Despite periodic improvements in output in forestry and in mineral fuels, real growth in the primary sector averages less than 1 per cent per annum over the projection horizon.

- Among the manufacturing sectors the outlook varies considerably.

  . Following a modest recovery, food and beverage industry growth recedes in step with slower population growth.

  . The textile and clothing industry recovers briskly in the mid-1980s only to decline in the early 1990s.

The Ontario Economy 1982 - 1995

TABLE 5
Growth rates by industry - Canada

		1961 -73	1974 -80	1981 -85	1986 -90	1991 -95
1.	Agriculture, Fishing	2.8	2.5	2.0	1.5	.5
2.	Forestry	5.5	-0.3	-0.8	2.5	1.4
3.	Mineral Fuel	10.7	-1.2	-0.2	1.7	4.0
4.	Other Mining	4.3	0.9	-2.2	1.2	.2
5.	Food & Beverages	4.3	2.0	1.2	1.4	.5
6.	Textiles, Clothing	6.7	0.8	1.2	2.1	-0.6
7.	Wood, Furniture	5.8	1.4	-0.6	2.4	1.6
8.	Paper & Printing	4.9	2.7	.5	2.8	1.5
9.	Metal Fabricating	6.0	1.8	.3	2.1	.6
10.	Motor Vehicles	8.7	-1.5	4.6	.9	1.0
11.	Machinery & Trans.	15.3	2.7	-1.2	3.3	.5
12.	Electrical Products	8.1	0.5	-0.3	1.6	-0.3
13.	Chemical, Petroleum	8.3	2.8	2.4	3.6	1.9
14.	Non-Metallic Mineral	5.9	-0.5	1.3	2.5	1.5
15.	Other Manufacturing	4.7	1.8	.8	1.0	-2.5
	Total Manufacturing	6.6	1.6	.9	2.2	.8
16.	Construction	4.5	2.4	2.1	2.3	2.5
17.	Utilities	8.0	5.2	3.6	4.5	3.8
	Total Goods	6.0	1.9	1.2	2.4	1.5
18.	Transportation	6.6	2.4	1.5	2.7	1.6
19.	Communication	6.9	8.3	5.9	7.2	6.1
20.	Trade	6.2	2.7	1.5	3.7	2.6
21.	Finance, Insurance & Real Estate	5.2	5.0	3.3	4.0	2.6
22.	Other Service Industries	6.0	5.6	4.0	4.4	3.6
	Total Service Industries	6.0	4.5	3.0	4.2	3.2
23.	Government Sector	3.7	-0.0	1.5	1.5	1.6
	TOTAL (GDP at Factor Cost)	5.6	2.9	2.2	3.3	2.4

- The wood and furniture industries reflect the depressed U.S. and Canadian housing sectors in the near term, but recover moderately in the decade following 1984.
- The 'metal fabricating', the 'machinery and other transportation equip-

Chart 3

Shares of Major Industry Groups in Total Output

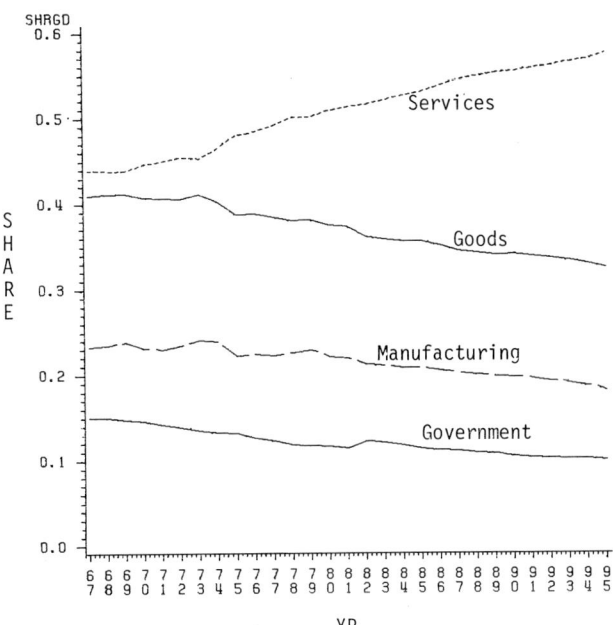

ment' and the 'electrical products' industries grow very slowly in the near term and again in the early 1990s after weak recoveries in the mid-1980s and 1990s. This reflects the general pattern of machinery and equipment investment spending and continued strong import competition.

- Motor vehicle output recovers from very low levels in the recent past, but grows slowly thereafter.

- The paper, printing and allied industries put in a steady performance throughout the outlook horizon.

- Two winners in the manufacturing sector are the chemical and petroleum products industry and the non-metallic mineral products industry. The former benefits from major new investment and from the assured feedstock supply during this period, while the latter benefits from major project construction in the later 1980s and the 1990s.

- The construction industry recovers only slowly from its current depressed state. However in 1985 and again in 1990, the construction industry experiences 9 plus per cent growth due to the assumed timing of major energy project activity.

- Utilities also suffer a decline in 1982 output, along with every other commercial sector except mineral fuels. However, recovery is swift in this sector, and strong growth persists into the mid-1990s. Again large capacity additions underlie this growth.

- All commercial service sector industries recover quickly from the 1982 downturn. Furthermore, all experience steady growth throughout the projection. The communications sector, in particular, sees fast growth, which is consistent with the expected 'information economy' and with past trends.

- The government sector is assumed to grow steadily at about 1½ per cent per annum. This conforms with present government policy and allows for a further gradual reduction in the relative size of government - a long established trend.

4
Provincial projection – PRISM

ASSUMPTIONS

Three other sets of assumptions or exogenously-projected variables are required to generate the provincial projections from the national industry detail. First are provincial output shares by industry. These are applied to the national industry output to calculate output by sector and province. The sum of industry outputs in each province yields, of course, an estimate of aggregate provincial real output. The second set of assumptions concerns differences across provinces in labour productivity in each sector. The third basic set of exogenous projections is for provincial population - especially the population of labour-force age.

It is important to note that these three sets of assumptions provide a check on one another. Together, the output-share and productivity assumptions will have a major impact on labour demand, while the population projections will affect labour supply. Too high or low a resulting unemployment rate for a province will signal an inconsistency on one or more of the sets of assumptions.

The three areas for which additional PRISM assumptions must be made are generally projected to reflect historical trends. Tables 6 through 8 show details of these assumptions.

\- Output Shares: In PRISM there are three sources of change in aggregate provincial output shares. These are:

. Changes in the composition of national final demand. These can change the industrial mix of national output and therefore one province's output share relative to others, even if provincial shares within sectors remain unchanged. A change in the industrial mix will affect most strongly those provinces specializing in the sectors affected. (E.g., a higher demand for cars raises Ontario's share of national output relative to the shares of other provinces, since Ontario is the major auto-producing province.)

- Changes in the provincial shares of primary and secondary sectors brought about by geographic shifts in investment, population, resource availability, population skills, shifts in incentives, etc. (E.g., east coast natural resource development or expansion of Alberta's petrochemical industry capacity.)

- Changes in provincial shares of 'nontransportable' or 'nontradeable' sectors - largely services - resulting from 1 and/or 2 above. (E.g., growth in retail trade or real estate agencies following an expansionary boom in a particular region.)

The first and third changes in output shares are for the most part determined within PRISM itself. The second set of changes requires industrial output shares to be adjusted to reflect specific knowledge about major investment programs like those enumerated in Table 2.

Additionally, for many sectors for which specific information on future shifts is unavailable, past trends - where strong and consistent - are extrapolated into the future. Charts 4, 6, 7 and 9 illustrate the extrapolations of share shifts made for 'mining' (other than mineral fuels), 'pulp and paper', 'primary metals and metal fabricating' and 'chemicals'. Charts 5 and 8 illustrate the pattern of mineral fuels and construction shares emerging from a combination of the major projects of Table 2 and judgemental projections based on past trends. The provincial shares of total output from this projection are provided in Table 6 (and plotted in Chart 14). Charts 10 and 11 show the evolution of output shares for manufacturing and total goods-producing.

- The Newfoundland and Nova Scotia shares of mineral fuels jump markedly as Sable Island and Hibernia reach production. The rising shares in the two eastern provinces shown in Chart 5 do reduce Alberta's share somewhat, as do increases in output of oil (and coal) in British Columbia,

Chart 4

Output Shares By Province - Other Mining

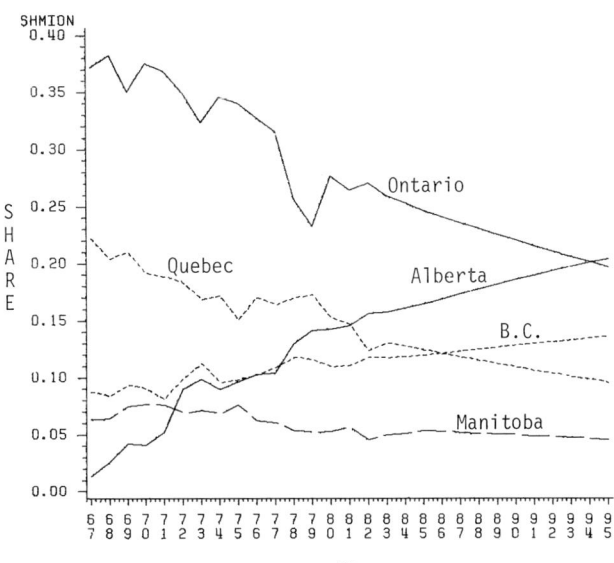

Chart 5

Output Shares By Province - Mineral Fuels

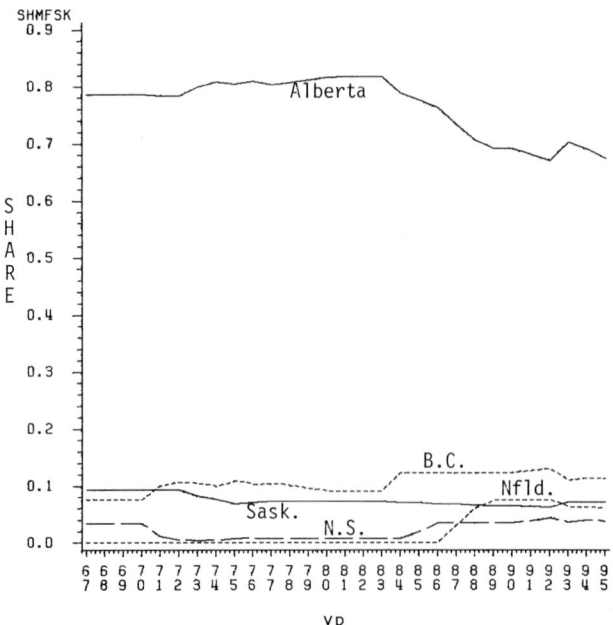

Chart 6

Output Shares By Province - Pulp and Paper

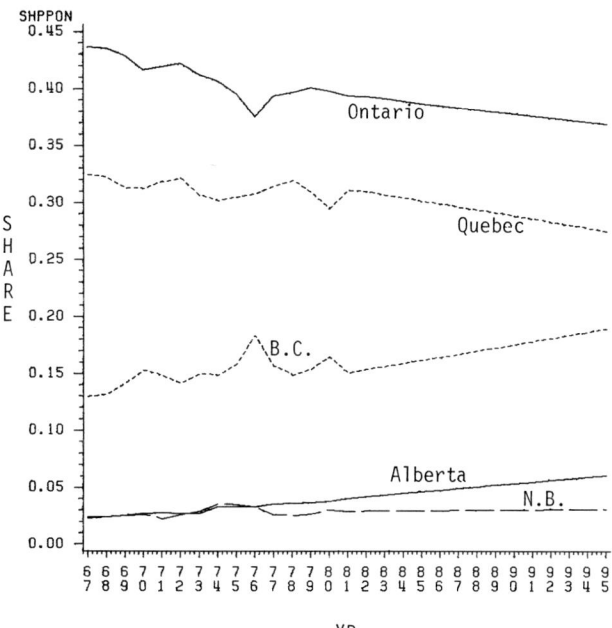

Chart 7

Output Shares by Province - Primary Metal and Metal Fabricating

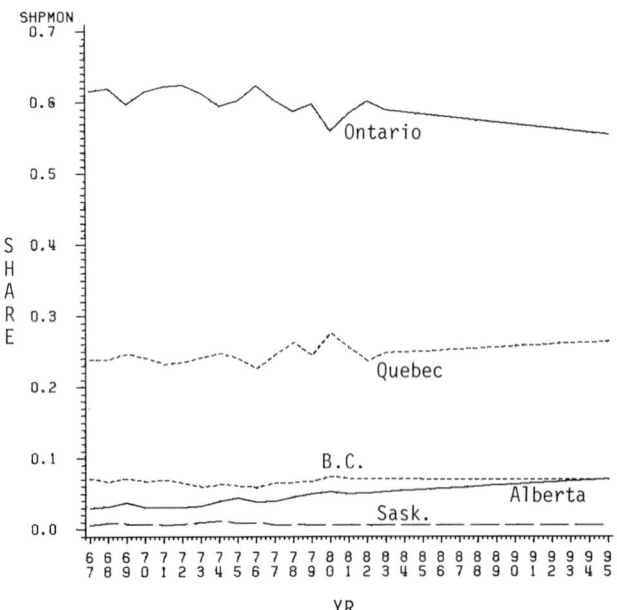

Chart 8

Output Shares By Province - Construction

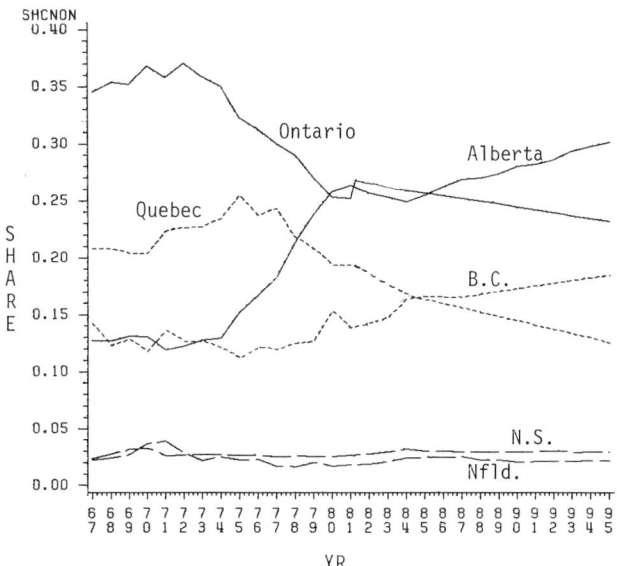

Chart 9

Output Shares By Province - Chemicals and Petroleum

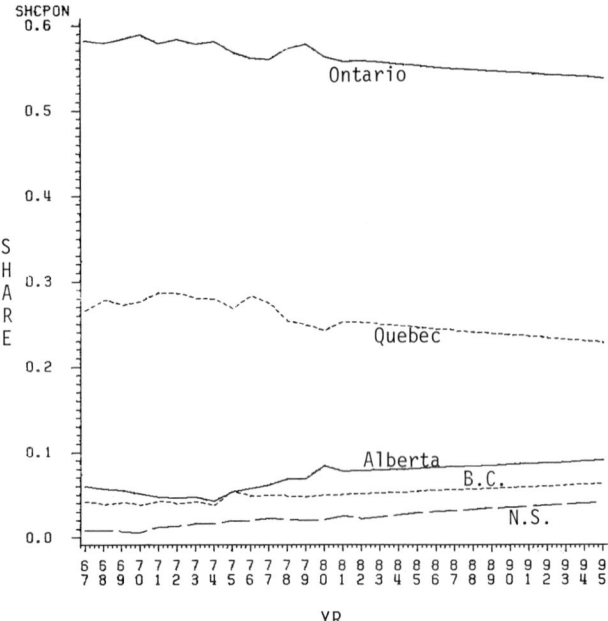

TABLE 6
Real output shares by province

	Annual averages				
	1967 -73	1974 -80	1981 -85	1986 -90	1991 -95
Newfoundland	.014	.014	.0142	.0150	.0155
Prince Edward Island	.003	.003	.0027	.0025	.0025
Nova Scotia	.026	.024	.0289	.0244	.0249
New Brunswick	.019	.019	.0186	.0184	.0184
Quebec	.239	.238	.228	.221	.215
Ontario	.419	.403	.390	.386	.380
Manitoba	.043	.040	.038	.037	.037
Saskatchewan	.038	.036	.036	.036	.037
Alberta	.088	.104	.122	.126	.132
British Columbia	.109	.116	.125	.130	.135
Territories	.002	.002	.0026	.0027	.0033

Saskatchewan and (in 1995) the Territories.

- Relative output shares have shifted radically in 'other mining' in the last decade and a half. In this scenario, these trends are projected to continue as shown in Chart 4.

- In the construction industry provincial shares have also moved significantly, albeit less smoothly than in some other sectors. Generally the trends are projected to continue although it could be argued that the most important regional shifts have now taken place and that a more stable period lies ahead. The continued high projected share for Alberta may thus be overly optimistic for that province.

- Productivity Differences: The second group of assumptions necessary to PRISM centres on provincial industrial labour productivity differentials. These variables affect, of course, the level and mix of provincial/industrial

Chart 10

Output Shares By Province - Manufacturing

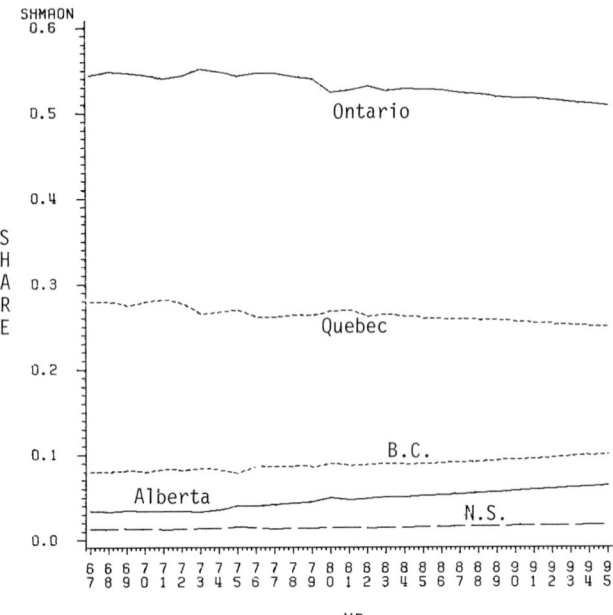

Chart 11

Output Shares By Province - Goods Producing

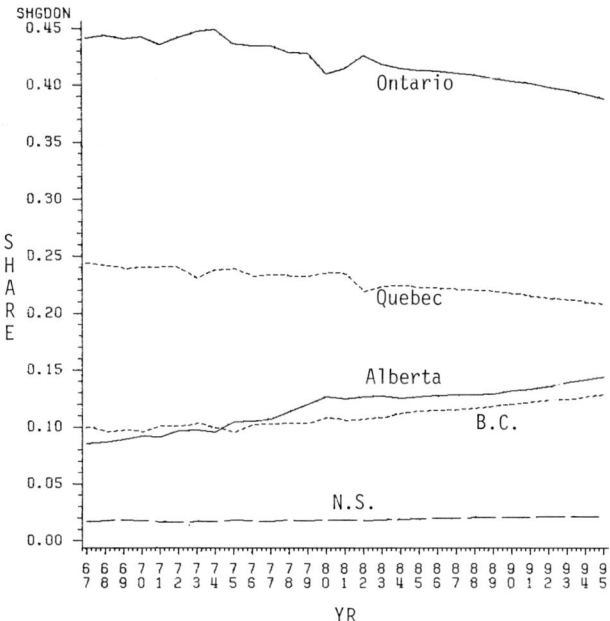

The Ontario Economy 1982 - 1995

TABLE 7
Labour productivity by province: all industries (per cent deviation from national)

	Annual averages				
	1967 -73	1974 -80	1981 -85	1986 -90	1991 -95
Newfoundland	-17	-16	-19	-18	-19
Prince Edward Island	-39	-37	-38	-39	-39
Nova Scotia	-19	-21	-20	-19	-19
New Brunswick	-22	-21	-23	-24	-24
Quebec	-12	-7	-6	-6	-6
Ontario	10	5	1	1	0
Manitoba	-9	-10	-12	-11	-10
Saskatchewan	-10	-10	-11	-10	-9
Alberta	11	16	20	20	20
British Columbia	7	7	10	10	10

employment. Again, historical levels and trends are the major determinants of the provincial productivity-differential projections.

Aggregate productivity does vary considerably from province to province as shown in Table 7, and for a variety of reasons, among which differences in industrial composition is a key element.

- Table 8, for example, shows that P.E.I. manufacturing productivity compares very unfavourably with the national average. This is due in part to the almost exclusive concentration of manufacturing in P.E.I. in the labour intensive food, feed and beverage industry.

- On the other hand, Alberta's strong showing in terms of manufacturing productivity reflects in part the relatively large share of the chemical and petroleum products and non-metallic minerals industries in Alberta; both are high productivity sectors nationally. Moreover, Alberta's

TABLE 8
Productivity - manufacturing (per cent deviation from national)

	Annual averages				
	1967 -73	1974 -80	1981 -85	1986 -90	1991 -95
Newfoundland	-32	-45	-55	-60	-64
Prince Edward Island	-37	-50	-51	-52	-54
Nova Scotia	-43	-39	-38	-35	-33
New Brunswick	-33	-32	-31	-32	-33
Quebec	-13	-12	-8	-6	-4
Ontario	15	13	9	7	5
Manitoba	-29	-21	-17	-12	-6
Saskatchewan	-8	-15	-12	-12	-12
Alberta	5	5	11	13	14
British Columbia	1	-1	7	10	12

attempt to develop its chemical industry on a world scale has added the efficiencies of large scale to its productivity results.

. An additional factor favouring relative productivity improvements in Alberta and British Columbia is the relatively newer vintage of their capital stocks. This relatively newer capital stock is more likely to embody recent technological innovation. The strong investment anticipated in the west suggests a continuation of these productivity trends.

. Ontario's large scale manufacturing facilities have dominated the national economy and accounted for the province's high relative productivity in the past. However, as new economic development has tended to shift west and Quebec's productivity has increasingly 'caught up' to the national, Ontario productivity has settled towards the national average.

Barring the unusual circumstances of Alberta and British Columbia where

TABLE 9
Productivity - goods producing (per cent deviation from national)

	Annual averages					
	1967 -73	1974 -80	1981 -85	1986 -90	1991 -95
Newfoundland	-11	-11	-20	-19	-21
Prince Edward Island	-60	-58	-62	-62	-61
Nova Scotia	-35	-33	-33	-29	-27
New Brunswick	-28	-25	-25	-25	-25
Quebec	-11	-8	-3	-1	3
Ontario	13	10	6	6	6
Manitoba	-23	-22	-22	-19	-14
Saskatchewan	-26	-28	-28	-27	-26
Alberta	21	18	12	7	1
British Columbia	6	3	7	6	4

the industrial composition is weighted towards high productivity industries, and where capital and management are on average newer and more productive, the tendency in recent years has been for aggregate provincial productivities, particularly in central Canada, to converge rather than drift apart.*

Over the forecast horizon these trends are expected to continue with the total relative productivity of most provinces moving little or moving towards the national average with the significant exception of Alberta and British Columbia. The productivity of the two western provinces is expected to increase relative to the national average because of the high proportion of total new investment going to these provinces and leading to industrial diversification following the development of the mineral fuels sector.

In the case of Newfoundland it is conceivable that development of the

---

* See Pierre Fortin, Economic Growth in Quebec (1978-80): The Human Capital Connection, PEAP Policy Study No. 82-4, September 1982.

mineral fuels sector would buoy relative productivity in at least some sectors as it did in the 1970s in Alberta and British Columbia. However the present projections adopt the 'safer' but more pessimistic course of continuing a relatively poor productivity-differential for Newfoundland into the future. Nova Scotia manufacturing does show an improvement in its differential through the projection; this, however, is more the result of a larger share of the high-productivity chemical and petroleum sector in its manufacturing mix than an improvement in its productivity differential within individual sectors.

Note finally that while past trends have generally been extended for individual sectors within provinces, the result for aggregate provincial productivity differentials also depends on the changing relative industrial mix within sectors. The aggregate results reported in Tables 7, 8 and 9, and in Chart 12 are thus not solely the result of sector-by-sector and province-by-province extrapolation. Note, for example, in Table 8 that the productivity differential of Alberta does continue to improve in the 1980s and 1990s reflecting the basic extrapolations discussed above. However, the aggregate differential for Alberta levels off in the 1980s and 1990s (see Table 7 and Chart 12). The reason is the high share of mineral fuels and the growing share of mining in Alberta - each of these sectors is projected nationally to feature low or negative productivity growth.

Population:

- Finally, the growth of total population and of population of labour-force age by province (see Chart 13 and Table 10) incorporates exogenous projections done independently by David Foot. (David Foot, Canada's Population Outlook, Canadian Institute for Economic Policy, Ottawa, 1982.) The basic scenario used was of 'high westward migration'; however, some reduction in the extent of net migration to Alberta, and out of Newfoundland, was incorporated in the scenario in order to achieve a consistent projection. Generally the labour force source population growth is expected to decelerate (see Table 10).

  . The slowest growth is expected from Quebec west to Saskatchewan.
  . The strongest growth occurs in Alberta and British Columbia which benefit from westward net migration.
  . A low rate of natural increase, as well as some out-migration, keeps

Chart 12

Productivity: Ratio to National Average

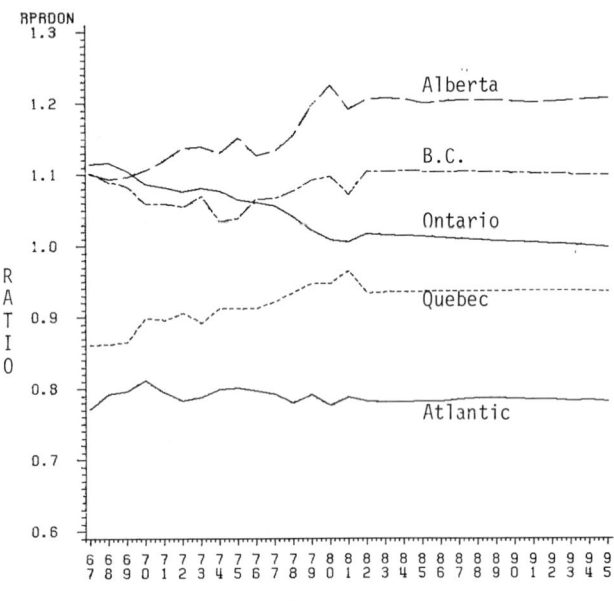

TABLE 10
Growth of labour-force source population

	Annual averages				
	1967 -73	1974 -80	1981 -85	1986 -90	1991 -95
Canada	2.7	2.1	1.4	1.0	.9
Newfoundland	2.4	2.3	2.0	1.5	1.3
Prince Edward Island	1.9	2.3	1.1	1.2	1.2
Nova Scotia	2.2	1.9	1.1	1.0	1.0
New Brunswick	2.2	2.3	1.3	1.2	1.2
Quebec	2.2	1.6	.5	.2	.3
Ontario	3.0	2.0	1.3	1.0	.9
Manitoba	1.6	1.1	.8	.8	.6
Saskatchewan	.3	1.8	1.6	.7	.5
Alberta	3.4	4.1	3.3	2.2	1.5
British Columbia	4.1	2.8	2.5	2.0	1.6

population growth low in Quebec and, to a lesser extent, in Manitoba (and, in the later 1980s, Saskatchewan). (Fertility fell in Quebec before the other provinces and the 'baby boom' was smaller and ended earlier.)

. For Newfoundland net out-migration is not as high as in some past estimates, reflecting a relative increase in opportunity in the province due to Hibernia and related developments.

Chart 13

Population By Region

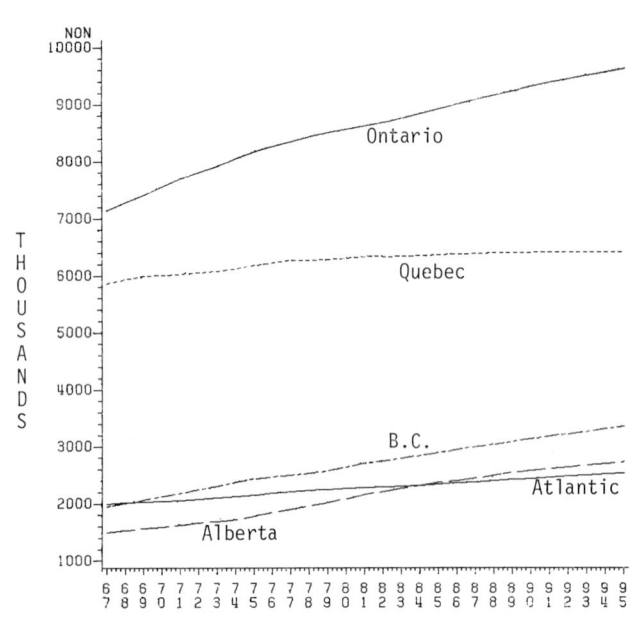

5
Provincial projection: overview of results

OUTPUT

Chart 14 illustrates the pattern of total output shares by region over the projection horizon. Generally, central Canada experiences a declining share of output while the shares of Alberta and British Columbia grow. This is not, of course, an indication of negative growth in central Canada, but rather a result of relatively strong growth in the west (see Table 11 and Chart 15). The output share of the Atlantic provinces remains virtually constant. Within the Atlantic region there are (small) increases in share for Newfoundland and Nova Scotia, and decreases for New Brunswick and Prince Edward Island.

OUTPUT DEFLATORS

Table 12 presents a summary of provincial output deflators, while Table 13 shows shares of nominal output by province. It is important to note that the output deflators are measures of the prices of the output of the provinces and not measures of inflation in the prices of consumer goods or other expenditure items. (No estimate has been made of consumer-price inflation by province, and indeed the differentials are often not large.) As an output price, a rise in this deflator is 'good' for the province concerned, in that it represents a shift in the terms of trade in its favour. Note that the highest growth rate for 1973-80 and 1981-85 is for Alberta, reflecting the increased relative prices for oil and gas.

. The continued move towards world prices in oil and gas keeps the

Chart 14

Share of Total Output - By Regions

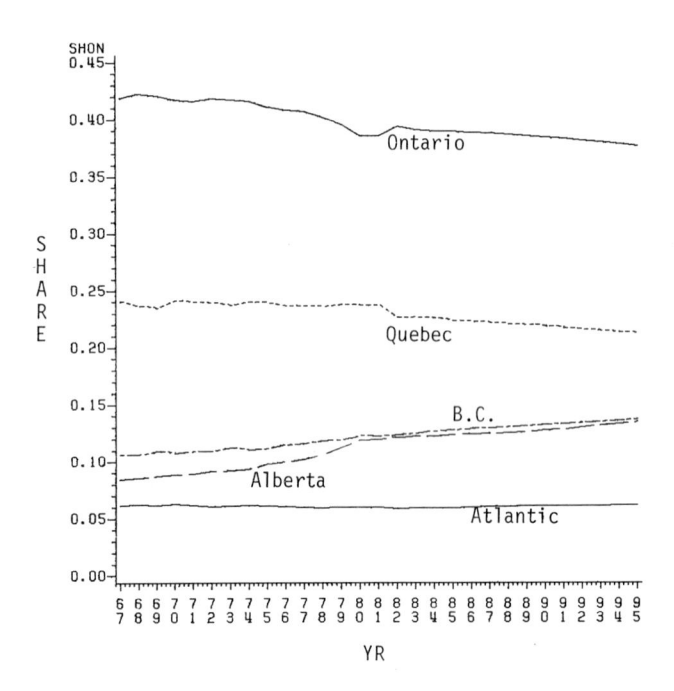

TABLE 11
Growth rates of real output by province

	Annual averages				
	1967 -73	1974 -80	1981 -85	1986 -90	1991 -95
Canada	5.2	2.9	2.2	3.3	2.4
Newfoundland	5.2	3.4	2.7	4.5	2.6
Prince Edward Island	5.2	2.4	.9	2.9	2.3
Nova Scotia	4.1	2.4	2.1	3.7	2.9
New Brunswick	5.5	2.2	2.1	3.2	2.5
Quebec	5.0	2.9	1.0	2.8	1.9
Ontario	5.1	1.8	2.4	3.0	2.0
Manitoba	4.7	1.6	1.8	3.1	2.2
Saskatchewan	1.5	3.3	2.1	3.7	3.0
Alberta	6.9	6.7	3.0	3.9	3.6
British Columbia	6.3	4.2	3.0	4.0	3.2
Territories	8.4	.7	4.7	4.5	7.6

Alberta deflator above the national average in 1981-85. In the later 1980s, oil prices move more at average rates, and the share of Alberta in mineral-fuels production declines, leaving its output deflator at about the national average.

- The deflators of Newfoundland and Nova Scotia rise in the 1980s, reflecting the increased weight of oil and gas production in their economies. There is a similar surge for the Territories in the early 1990s.

- Quebec and Ontario remain somewhat below the national average in the 1980s, as in 1973-80. The 'terms of trade' shift gradually against each as oil and gas consumers rather than producers.

Shares of nominal output (Table 13) reflect not only real-output but also

Chart 15

Output Growth - By Region

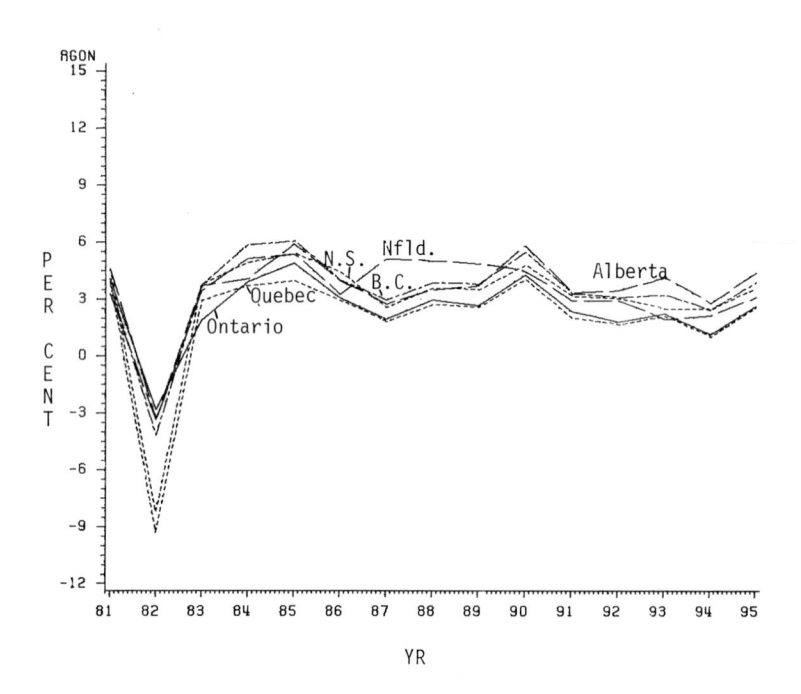

Provincial Projection: Overview of Results

TABLE 12
Growth rates of GDP deflators by province

	Annual averages				
	1967 -73	1974 -80	1981 -85	1986 -90	1991 -95
Canada	5.1	10.7	10.1	9.1	8.9
Newfoundland	4.7	10.5	10.9	14.0	8.8
Prince Edward Island	6.5	10.9	9.2	9.8	9.4
Nova Scotia	6.2	9.9	11.1	10.3	9.5
New Brunswick	5.1	11.2	9.3	9.0	8.7
Quebec	4.0	10.1	9.9	8.9	8.5
Ontario	5.3	10.1	9.9	8.9	8.6
Manitoba	5.2	9.9	9.8	9.1	8.6
Saskatchewan	5.7	11.7	9.2	9.2	8.9
Alberta	5.1	12.7	10.9	9.2	9.0
British Columbia	6.1	9.5	10.2	9.3	8.7
Territories	6.4	13.6	8.8	9.3	14.6

terms-of-trade shifts, and are thus a first measure of provincial income.

- Note the major increases in share for Alberta over both the 1970s and 1980s, the combined result of high activity and a favourable terms-of-trade shift.
- Note also the rise in share in the late 1980s for Newfoundland and Nova Scotia.
- A long-term loss of income share is concentrated in the four central provinces - New Brunswick through Manitoba.

EMPLOYMENT; UNEMPLOYMENT RATES

Table 14 and Chart 16 show average unemployment rates by province and region respectively.

The Ontario Economy 1982 - 1995

TABLE 13
Shares of nominal GDP - by province

	Annual averages				
	1967 -73	1974 -80	1981 -85	1986 -90	1991 -95
Newfoundland	.0138	.0137	.0140	.0174	.0192
Prince Edward Island	.0028	.0029	.0028	.0027	.0027
Nova Scotia	.0254	.0245	.0236	.0257	.0271
New Brunswick	.0193	.0199	.0190	.0185	.0184
Quebec	.243	.231	.216	.206	.199
Ontario	.412	.382	.369	.361	.351
Manitoba	.043	.040	.036	.036	.035
Saskatchewan	.039	.043	.042	.042	.043
Alberta	.089	.122	.140	.155	.163
British Columbia	.109	.118	.124	.131	.136
Territories	.0028	.0033	.0035	.0037	.0049

- Unemployment rates rise in all provinces in 1981-85, relative to 1973-80. The most notable increases are in Quebec (4 percentage points) and Alberta (2+ percentage points but a 50 per cent increase). There is also a large relative increase in Ontario, but the province does stay below the national average.
- In the later 1980s and early 1990s the national unemployment rate falls, and all provincial rates as well - but to varying degrees.
- The decline in the Newfoundland rate is still such as to leave it with the highest rate - despite Hibernia activity and a high growth rate for employment. The unemployment rate remains high because increased activity encourages many 'discouraged workers' to re-enter the labour force. Even so, the Newfoundland participation rate remains the lowest in the country.
- The Quebec unemployment rate falls more than the national average in

Provincial Projection: Overview of Results

TABLE 14
Unemployment rates - by province

	Annual averages				
	1967 -73	1974 -80	1981 -85	1986 -90	1991 -95
Canada	5.2	7.3	10.3	9.0	7.4
Newfoundland	7.9	14.5	16.2	14.4	12.3
Prince Edward Island	7.3	9.5	12.8	12.3	11.5
Nova Scotia	5.8	9.3	12.4	11.0	9.5
New Brunswick	6.5	10.9	14.1	13.2	11.4
Quebec	6.4	9.1	13.2	11.8	9.8
Ontario	4.2	6.4	9.4	7.9	6.3
Manitoba	4.4	5.2	7.3	6.2	5.8
Saskatchewan	3.3	3.9	5.8	5.2	4.1
Alberta	4.4	4.0	6.6	6.2	4.5
British Columbia	6.5	7.8	10.5	9.2	7.8

the later 1980s and early 1990s even with a considerable increase in participation. The low rate of labour-force growth in Quebec should give a downward thrust to the Quebec unemployment rate in the later 1980s and 1990s.

. Unemployment rates also fall in the west, but not to earlier low levels. Of course, western rates are especially sensitive to assumptions about interprovincial migration.

PERSONAL INCOME

All incomes are deflated to real terms by the national CPI. Since estimation of personal income is only rudimentary as yet in the PRISM model, the figures should be taken as suggestive only. Real personal income per capita is summarized in Table 15 and Chart 17 and is expressed as a ratio to the national average. Growth rates of real personal disposable income per capita

Chart 16

Unemployment Rates - By Region

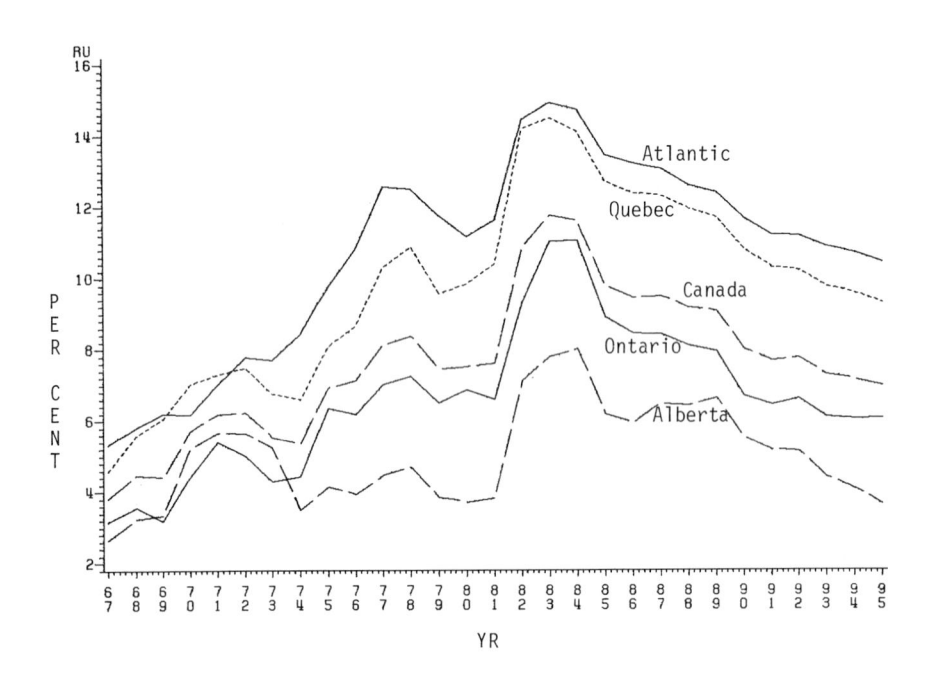

TABLE 15
Ratio of real disposable income per capita to national

	Annual averages				
	1967-73	1974-80	1981-85	1986-90	1991-95
Newfoundland	.66	.69	.68	.73	.76
Prince Edward Island	.69	.73	.76	.74	.75
Nova Scotia	.80	.81	.81	.82	.84
New Brunswick	.74	.76	.74	.74	.75
Quebec	.90	.91	.91	.91	.92
Ontario	1.15	1.09	1.07	1.05	1.02
Manitoba	.96	.97	.95	.95	.95
Saskatchewan	.85	.99	.97	1.01	1.07
Alberta	1.00	1.05	1.13	1.16	1.20
British Columbia	1.09	1.10	1.10	1.10	1.10

are summarized in Table 16.

- In 1981-85 three provinces - Ontario, Quebec and British Columbia - are below the low national average growth rate. With the higher portions of income coming from transfers in the Atlantic provinces (and, to some extent, Quebec), declines in output and activity have less of a relative impact on incomes. In the west, the fact that real growth itself is above average keeps income growth up.

- In 1986-90 income growth is above average in Newfoundland and Nova Scotia - again reflecting energy income. This assumes that some of the additional revenue from royalties is passed on to residents by their provincial governments in the form of reduced tax burdens.

- Alberta recovers its former position with respect to income growth, and Saskatchewan also does well. The British Columbia per capita income is in part kept near the national average by in-migration.

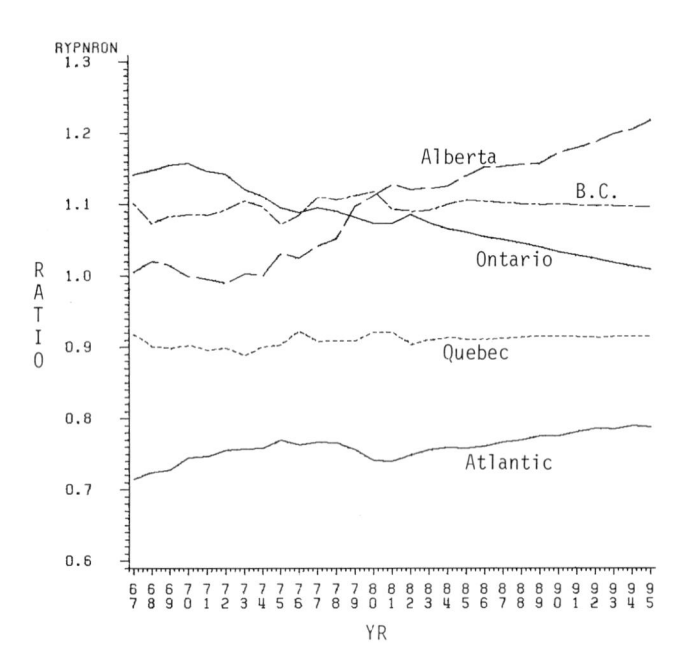

Chart 17

Real Personal Disposable Income Per Capita
Ratio to National Average

Provincial Projection: Overview of Results

TABLE 16
Growth rates of real personal disposal income per capita - by province

	Annual averages				
	1967 -73	1974 -80	1981 -85	1986 -90	1991 -95
Canada	4.3	2.5	.8	1.4	1.3
Newfoundland	5.4	2.1	2.2	2.6	1.7
Prince Edward Island	7.0	2.4	.7	1.4	1.5
Nova Scotia	5.3	2.5	1.0	1.8	1.8
New Brunswick	5.5	2.0	1.1	1.4	1.4
Quebec	4.0	3.1	.6	1.5	1.3
Ontario	4.0	1.9	.6	.9	.8
Manitoba	5.3	1.6	1.1	1.5	1.2
Saskatchewan	4.5	2.2	1.5	2.6	2.3
Alberta	4.0	4.1	1.3	2.0	2.1
British Columbia	4.2	2.7	.6	1.3	1.3

- The rate of growth for Saskatchewan may be a trifle high as a result of a combination of good returns to farmers and relatively low population growth.

- A comparison of the later 1980s and early 1990s and 1962-73 shows that personal-income differentials have generally narrowed and are projected to continue to do so. Exceptions are Alberta and Saskatchewan - which continue to rise above the national average, and New Brunswick and Manitoba - which neither gain nor lose.

- Ontario, Alberta and British Columbia remain the principal 'have' provinces. Alberta supplants Ontario and British Columbia as the province of highest relative income - but the ratio is not much greater than that of Ontario in 1967-73.

- Newfoundland catches up to New Brunswick and Prince Edward Island, especially with Hibernia. Still, the Atlantic remains well behind the central and western regions.

6
Results by province

Tables 17 and 18 provide summary results by province. For the very near term it should be noted that because provincial output data are published only after considerable delay, available labour-force survey data have been used to give rough estimates of current output and employment developments.

The Newfoundland economy finally begins to enjoy a taste of prosperity in this scenario as a result of the Hibernia oil field development.

- RDP grows in excess of 3 per cent per annum through the 1980s after a 3½ per cent decline in 1982.

    . Construction RDP benefits from the Hibernia development in the 1980s.
    . Utilities grow robustly.
    . Associated services grow briskly.

- Employment grows very fast, particularly during the building phase of Hibernia.

    . The unemployment rate winds down from over 17 per cent in 1983 and 1984 to under 12 per cent by 1995.

- Real personal disposable income per capita grows rapidly but only lifts Newfoundlanders' income from 2/3 to 3/4 of the national average.

Prince Edward Island experiences a severe recession in 1982 suffering a

INSTITUTE FOR POLICY ANALYSIS, UNIVERSITY OF TORONTO
PRISM82A *** REVISED LO-TREND BASE CASE *** SEPT/82

Table 17

REAL GDP

(MILLIONS OF 1971 DOLLARS)

	1981	1982	1983	1984	1985	1986	1987	1988
CANADA	119693	113754	116850	121687	127691	131909	134864	139140
NEWFOUNDLAND	1669	1611	1667	1752	1845	1905	2002	2101
PRINCE EDWARD ISLAND	337	302	308	317	327	336	343	353
NOVA SCOTIA	2912	2671	2772	2908	3065	3200	3281	3398
NEW BRUNSWICK	2214	2118	2175	2258	2359	2431	2484	2559
QUEBEC	28383	25742	26487	27467	28553	29379	29911	30719
ONTARIO	46118	44807	45646	47424	49736	51264	52270	53807
MANITOBA	4552	4274	4392	4554	4757	4908	5013	5160
SASKATCHEWAN	4239	4059	4193	4349	4566	4718	4836	5038
ALBERTA	14275	13798	14308	14887	15761	16384	16833	17421
BRITISH COLUMBIA	14636	14022	14544	15395	16325	16974	17473	18143
YUKON & N.W.T.	302	294	304	319	341	351	359	380
EXTERNAL & DISCREPANCY	56	56	56	56	56	56	56	56

GDP DEFLATORS

(1971 = 1.0)

	1981	1982	1983	1984	1985	1986	1987	1988
CANADA	2.51	2.82	3.10	3.40	3.77	4.13	4.53	4.95
NEWFOUNDLAND	2.45	2.76	3.04	3.37	3.78	4.17	4.98	5.88
PRINCE EDWARD ISLAND	2.67	2.96	3.23	3.54	3.93	4.32	4.77	5.24
NOVA SCOTIA	2.47	2.76	3.04	3.35	3.81	4.30	4.75	5.22
NEW BRUNSWICK	2.61	2.90	3.16	3.45	3.81	4.16	4.55	4.96
QUEBEC	2.38	2.66	2.92	3.21	3.56	3.88	4.25	4.63
ONTARIO	2.39	2.67	2.93	3.21	3.55	3.88	4.24	4.63
MANITOBA	2.46	2.73	3.00	3.29	3.66	4.00	4.38	4.79
SASKATCHEWAN	2.96	3.31	3.62	3.95	4.38	4.79	5.27	5.74
ALBERTA	3.06	3.49	3.86	4.22	4.64	5.12	5.62	6.11
BRITISH COLUMBIA	2.49	2.79	3.06	3.40	3.78	4.15	4.56	4.98
YUKON & N.W.T.	3.49	3.89	4.21	4.57	5.05	5.52	6.06	6.60

INSTITUTE FOR POLICY ANALYSIS, UNIVERSITY OF TORONTO
PRISM82A *** REVISED LO-TREND BASE CASE *** SEPT/82

Table 17 (cont'd.)
REAL GDP

(MILLIONS OF 1971 DOLLARS)

	1989	1990	1991	1992	1993	1994	1995
CANADA	143323	149935	153799	157290	161431	164043	169189
NEWFOUNDLAND	2203	2301	2369	2438	2485	2538	2616
PRINCE EDWARD ISLAND	363	376	385	393	403	410	422
NOVA SCOTIA	3516	3683	3799	3913	4010	4108	4254
NEW BRUNSWICK	2637	2756	2825	2891	2970	3024	3120
QUEBEC	31500	32767	33414	33965	34673	35019	35920
ONTARIO	55239	57594	58937	60008	61355	62053	63676
MANITOBA	5305	5531	5656	5769	5909	5989	6156
SASKATCHEWAN	5213	5480	5641	5799	5977	6112	6337
ALBERTA	18064	19114	19752	20429	21270	21860	22818
BRITISH COLUMBIA	18824	19848	20498	21130	21807	22335	23201
YUKON & N.W.T.	396	424	467	494	512	537	610
EXTERNAL & DISCREPANCY	56	56	56	56	56	56	56

GDP DEFLATORS

(1971 = 1.0)

	1989	1990	1991	1992	1993	1994	1995
CANADA	5.37	5.85	6.36	6.90	7.49	8.15	8.92
NEWFOUNDLAND	6.62	7.26	7.94	8.70	9.24	10.12	11.07
PRINCE EDWARD ISLAND	5.72	6.26	6.87	7.48	8.19	8.97	9.84
NOVA SCOTIA	5.69	6.22	6.85	7.52	8.11	8.92	9.77
NEW BRUNSWICK	5.38	5.85	6.37	6.88	7.48	8.12	8.87
QUEBEC	5.01	5.45	5.92	6.39	6.93	7.52	8.21
ONTARIO	5.01	5.45	5.91	6.39	6.93	7.53	8.22
MANITOBA	5.19	5.66	6.15	6.64	7.22	7.84	8.55
SASKATCHEWAN	6.22	6.78	7.37	7.97	8.74	9.51	10.38
ALBERTA	6.61	7.20	7.83	8.52	9.33	10.14	11.09
BRITISH COLUMBIA	5.40	5.89	6.42	6.96	7.51	8.17	8.94
YUKON & N.W.T.	7.18	7.86	8.85	9.94	10.60	11.92	15.40

INSTITUTE FOR POLICY ANALYSIS, UNIVERSITY OF TORONTO
PRISM82A *** REVISED LO-TREND BASE CASE *** SEPT/82

Table 17 (cont'd.)
REAL GDP

(YEAR-OVER-YEAR PER-CENT CHANGE)

	1981	1982	1983	1984	1985	1986	1987	1988
CANADA	4.03	-4.96	2.72	4.14	4.93	3.30	2.24	3.17
NEWFOUNDLAND	3.26	-3.44	3.44	5.12	5.33	3.21	5.09	4.98
PRINCE EDWARD ISLAND	6.92	-10.55	2.07	3.06	3.09	2.74	2.22	2.79
NOVA SCOTIA	4.56	-8.26	3.77	4.92	5.38	4.42	2.54	3.54
NEW BRUNSWICK	3.74	-4.35	2.71	3.81	4.48	3.03	2.17	3.03
QUEBEC	3.94	-9.31	2.90	3.70	3.95	2.90	1.81	2.70
ONTARIO	4.06	-2.84	1.87	3.89	4.88	3.07	1.96	2.94
MANITOBA	4.41	-6.11	2.75	3.70	4.45	3.17	2.14	2.95
SASKATCHEWAN	2.48	-4.25	3.31	3.73	4.98	3.32	2.51	4.18
ALBERTA	4.58	-3.34	3.69	4.05	5.87	3.95	2.74	3.49
BRITISH COLUMBIA	3.72	-4.20	3.72	5.85	6.04	3.98	2.94	3.83
YUKON & N.W.T.	11.05	-2.50	3.17	4.96	6.85	3.21	2.05	6.00
EXTERNAL & DISCREPANCY	-3.29	0.0	0.0	0.0	0.0	0.0	0.0	0.0

GDP DEFLATORS

(YEAR-OVER-YEAR PER-CENT CHANGE)

	1981	1982	1983	1984	1985	1986	1987	1988
CANADA	7.74	12.12	9.96	9.80	10.76	9.64	9.72	9.25
NEWFOUNDLAND	8.86	12.49	10.32	10.68	12.31	10.28	19.36	18.06
PRINCE EDWARD ISLAND	5.58	10.72	9.01	9.81	10.91	10.04	10.36	9.78
NOVA SCOTIA	9.43	12.03	9.86	10.44	13.53	12.91	10.43	9.88
NEW BRUNSWICK	6.56	11.37	8.88	9.24	10.42	9.21	9.38	8.97
QUEBEC	7.19	11.60	8.83	9.90	10.75	9.25	9.32	9.03
ONTARIO	7.59	11.65	9.82	9.56	10.63	9.26	9.38	9.09
MANITOBA	7.05	11.22	9.77	9.87	11.07	9.37	9.59	9.18
SASKATCHEWAN	5.02	11.87	9.35	9.28	10.64	9.57	9.96	8.88
ALBERTA	10.25	14.33	10.62	9.24	9.94	10.28	9.79	8.81
BRITISH COLUMBIA	7.23	11.70	9.94	11.10	11.11	9.81	9.74	9.25
YUKON & N.W.T.	5.10	11.55	8.33	8.49	10.39	9.51	9.73	8.92

INSTITUTE FOR POLICY ANALYSIS, UNIVERSITY OF TORONTO
PRISM82A *** REVISED LO-TREND BASE CASE *** SEPT/82

Table 17 (cont'd.)

REAL GDP
(YEAR-OVER-YEAR PER-CENT CHANGE)

	1989	1990	1991	1992	1993	1994	1995
CANADA	3.01	4.61	2.58	2.27	2.63	1.62	3.14
NEWFOUNDLAND	4.84	4.46	2.93	2.92	1.94	2.12	3.08
PRINCE EDWARD ISLAND	2.80	3.75	2.22	2.07	2.52	1.75	3.01
NOVA SCOTIA	3.47	4.76	3.14	3.01	2.49	2.43	3.55
NEW BRUNSWICK	3.04	4.51	2.52	2.33	2.75	1.81	3.16
QUEBEC	2.54	4.02	1.97	1.65	2.08	1.00	2.57
ONTARIO	2.66	4.26	2.33	1.82	2.24	1.14	2.61
MANITOBA	2.81	4.26	2.26	1.99	2.43	1.35	2.78
SASKATCHEWAN	3.47	5.11	2.95	2.80	3.05	2.26	3.69
ALBERTA	3.69	5.81	3.33	3.43	4.12	2.77	4.38
BRITISH COLUMBIA	3.76	5.44	3.28	3.08	3.20	2.42	3.88
YUKON & N.W.T.	4.27	6.90	10.18	5.92	3.52	4.99	13.51
EXTERNAL & DISCREPANCY	0.0	0.0	0.0	0.0	0.0	0.0	0.0

GDP DEFLATORS
(YEAR-OVER-YEAR PER-CENT CHANGE)

	1989	1990	1991	1992	1993	1994	1995
CANADA	8.46	8.88	8.80	8.36	8.63	8.81	9.43
NEWFOUNDLAND	12.63	9.64	9.38	9.55	6.29	9.48	9.40
PRINCE EDWARD ISLAND	9.18	9.56	9.62	8.91	9.46	9.54	9.70
NOVA SCOTIA	9.03	9.42	10.09	9.74	7.83	10.06	9.53
NEW BRUNSWICK	8.34	8.85	8.85	8.08	8.58	8.65	9.20
QUEBEC	8.28	8.67	8.63	7.93	8.50	8.54	9.12
ONTARIO	8.33	8.64	8.57	8.00	8.52	8.59	9.20
MANITOBA	8.47	8.94	8.76	7.98	8.62	8.65	9.08
SASKATCHEWAN	8.33	9.04	8.70	8.16	9.69	8.76	9.11
ALBERTA	8.13	8.96	8.65	8.85	9.45	8.74	9.35
BRITISH COLUMBIA	8.47	9.02	9.04	8.43	7.86	8.86	9.34
YUKON & N.W.T.	8.72	9.50	12.61	12.32	6.56	12.48	29.20

INSTITUTE FOR POLICY ANALYSIS, UNIVERSITY OF TORONTO
PRISM82A *** REVISED LO-TREND BASE CASE *** SEPT/82

Table 18
EMPLOYMENT
(THOUSANDS)

	1981	1982	1983	1984	1985	1986	1987	1988
CANADA	10933	10584	10616	10872	11345	11631	11853	12043
NEWFOUNDLAND	187	183	186	193	203	209	216	221
PRINCE EDWARD ISLAND	48	45	46	46	48	48	49	50
NOVA SCOTIA	330	312	315	325	339	349	356	363
NEW BRUNSWICK	262	256	258	263	274	281	286	291
QUEBEC	2685	2564	2571	2622	2709	2767	2808	2839
ONTARIO	4186	4096	4083	4174	4358	4465	4547	4616
MANITOBA	462	456	456	464	480	491	498	504
SASKATCHEWAN	432	426	428	436	454	464	473	484
ALBERTA	1093	1065	1077	1104	1167	1202	1228	1252
BRITISH COLUMBIA	1247	1181	1196	1245	1314	1356	1391	1423

UNEMPLOYMENT RATES
(PER CENT)

	1981	1982	1983	1984	1985	1986	1987	1988
CANADA	7.61	10.84	11.76	11.59	9.76	9.46	9.50	9.18
NEWFOUNDLAND	14.16	16.38	17.36	17.18	15.91	15.60	15.14	14.31
PRINCE EDWARD ISLAND	11.12	13.27	13.42	13.52	12.64	12.81	12.75	12.23
NOVA SCOTIA	10.17	13.56	13.47	12.93	11.63	11.34	11.40	11.05
NEW BRUNSWICK	11.64	14.33	15.13	15.25	13.96	13.77	13.60	13.24
QUEBEC	10.41	14.19	14.48	14.09	12.68	12.36	12.29	11.94
ONTARIO	6.60	9.30	11.03	11.06	8.90	8.45	8.43	8.10
MANITOBA	6.08	7.76	8.19	8.08	6.37	6.50	6.53	6.19
SASKATCHEWAN	4.63	5.96	6.27	6.48	5.55	5.45	5.64	5.18
ALBERTA	3.82	7.11	7.81	8.02	6.17	5.93	6.47	6.43
BRITISH COLUMBIA	6.72	12.02	12.57	11.77	9.48	9.45	9.58	9.35

INSTITUTE FOR POLICY ANALYSIS, UNIVERSITY OF TORONTO
PRISM82A *** REVISED LO-TREND BASE CASE *** SEPT/82

Table 18 (cont'd.)

EMPLOYMENT

(THOUSANDS)

	1989	1990	1991	1992	1993	1994	1995
CANADA	12224	12619	12759	12927	13219	13435	13639
NEWFOUNDLAND	228	236	241	246	252	258	263
PRINCE EDWARD ISLAND	51	52	53	53	54	55	56
NOVA SCOTIA	370	382	388	395	404	412	420
NEW BRUNSWICK	296	306	309	314	322	328	334
QUEBEC	2867	2941	2957	2978	3028	3061	3091
ONTARIO	4676	4816	4863	4914	5013	5081	5143
MANITOBA	509	522	525	529	539	545	550
SASKATCHEWAN	492	510	516	524	536	547	557
ALBERTA	1280	1338	1364	1396	1446	1485	1524
BRITISH COLUMBIA	1455	1515	1544	1577	1624	1663	1701

UNEMPLOYMENT RATES

(PER CENT)

	1989	1990	1991	1992	1993	1994	1995
CANADA	9.07	8.02	7.68	7.77	7.28	7.14	6.98
NEWFOUNDLAND	13.87	13.22	12.65	12.53	12.27	12.07	11.80
PRINCE EDWARD ISLAND	12.20	11.72	11.40	11.79	11.54	11.50	11.34
NOVA SCOTIA	10.91	10.29	9.80	9.73	9.54	9.40	9.20
NEW BRUNSWICK	13.14	12.19	11.81	11.83	11.33	11.10	10.79
QUEBEC	11.70	10.79	10.30	10.23	9.76	9.57	9.30
ONTARIO	7.95	6.69	6.45	6.62	6.10	6.05	6.07
MANITOBA	6.37	5.62	5.56	6.11	5.78	5.78	5.72
SASKATCHEWAN	5.07	4.55	4.14	4.26	4.13	4.05	3.89
ALBERTA	6.64	5.51	5.16	5.12	4.42	4.08	3.65
BRITISH COLUMBIA	9.36	8.25	8.01	8.21	7.74	7.56	7.37

INSTITUTE FOR POLICY ANALYSIS, UNIVERSITY OF TORONTO
PRISM82A *** REVISED LO-TREND BASE CASE *** SEPT/82

Table 18 (cont'd.)

EMPLOYMENT
(YEAR-OVER-YEAR PER-CENT CHANGE)

	1981	1982	1983	1984	1985	1986	1987	1988
CANADA	2.59	-3.19	0.30	2.41	4.36	2.52	1.91	1.60
NEWFOUNDLAND	1.39	-2.05	1.42	3.81	5.05	2.98	3.49	2.35
PRINCE EDWARD ISLAND	-0.53	-4.77	0.06	1.62	2.76	2.06	1.88	1.39
NOVA SCOTIA	0.63	-5.40	1.13	3.03	4.22	2.98	2.08	1.94
NEW BRUNSWICK	2.04	-2.06	0.47	2.21	4.04	2.48	2.04	1.65
QUEBEC	0.65	-4.52	0.30	1.96	3.34	2.13	1.48	1.10
ONTARIO	2.93	-2.16	-0.32	2.23	4.42	2.45	1.84	1.52
MANITOBA	0.74	-1.41	0.06	1.74	3.53	2.10	1.59	1.14
SASKATCHEWAN	2.04	-1.43	0.64	1.79	4.04	2.23	1.90	2.40
ALBERTA	5.97	-2.63	1.12	2.55	5.67	3.00	2.22	1.94
BRITISH COLUMBIA	4.71	-5.28	1.27	4.02	5.61	3.17	2.56	2.29

UNEMPLOYMENT RATES
(PER CENT)
(YEAR-TO-YEAR CHANGE)

	1981	1982	1983	1984	1985	1986	1987	1988
CANADA	0.09	3.23	0.92	-0.17	-1.83	-0.30	0.04	-0.32
NEWFOUNDLAND	0.62	2.22	0.99	-0.18	-1.27	-0.31	-0.46	-0.83
PRINCE EDWARD ISLAND	0.29	2.15	0.16	0.10	-0.88	0.16	-0.06	-0.51
NOVA SCOTIA	0.40	3.39	-0.10	-0.54	-1.30	-0.29	-0.05	-0.35
NEW BRUNSWICK	0.43	2.69	0.80	0.13	-1.29	-0.19	-0.17	-0.36
QUEBEC	0.56	3.78	0.29	-0.39	-1.41	-0.31	-0.07	-0.36
ONTARIO	-0.27	2.70	1.73	0.03	-2.16	-0.45	-0.02	-0.34
MANITOBA	0.57	1.69	0.42	-0.10	-1.71	0.14	0.03	-0.34
SASKATCHEWAN	0.21	1.33	0.31	0.21	-0.92	-0.10	0.18	-0.46
ALBERTA	0.12	3.29	0.69	0.22	-1.85	-0.24	0.54	-0.05
BRITISH COLUMBIA	-0.04	5.30	0.55	-0.81	-2.29	-0.03	0.13	-0.23

INSTITUTE FOR POLICY ANALYSIS, UNIVERSITY OF TORONTO
PRISM82A *** REVISED LO-TREND BASE CASE *** SEPT/82

Table 18 (cont'd.)

EMPLOYMENT

(YEAR-OVER-YEAR PER-CENT CHANGE)

	1989	1990	1991	1992	1993	1994	1995
CANADA	1.50	3.23	1.11	1.31	2.26	1.64	1.51
NEWFOUNDLAND	2.99	3.61	1.90	2.17	2.57	2.29	1.96
PRINCE EDWARD ISLAND	1.39	2.54	0.82	1.13	2.14	1.59	1.47
NOVA SCOTIA	1.91	3.33	1.51	1.80	2.23	2.10	1.97
NEW BRUNSWICK	1.66	3.25	1.23	1.52	2.47	1.88	1.81
QUEBEC	0.99	2.59	0.53	0.72	1.69	1.07	1.00
ONTARIO	1.30	2.99	0.98	1.05	2.02	1.36	1.20
MANITOBA	1.03	2.54	0.54	0.79	1.78	1.15	1.02
SASKATCHEWAN	1.76	3.47	1.31	1.59	2.25	1.97	1.79
ALBERTA	2.20	4.59	1.91	2.35	3.61	2.66	2.62
BRITISH COLUMBIA	2.26	4.15	1.88	2.14	2.96	2.44	2.27

UNEMPLOYMENT RATES

(PER CENT)
(YEAR-TO-YEAR CHANGE)

	1989	1990	1991	1992	1993	1994	1995
CANADA	-0.11	-1.05	-0.33	0.08	-0.48	-0.14	-0.16
NEWFOUNDLAND	-0.44	-0.65	-0.57	-0.12	-0.26	-0.21	-0.27
PRINCE EDWARD ISLAND	-0.03	-0.49	-0.32	0.39	-0.24	-0.04	-0.16
NOVA SCOTIA	-0.14	-0.62	-0.49	-0.06	-0.20	-0.14	-0.20
NEW BRUNSWICK	-0.10	-0.95	-0.38	0.02	-0.49	-0.23	-0.32
QUEBEC	-0.24	-0.91	-0.48	-0.07	-0.46	-0.19	-0.28
ONTARIO	-0.15	-1.25	-0.24	0.17	-0.52	-0.05	0.01
MANITOBA	-0.18	-0.75	-0.06	0.55	-0.33	0.00	-0.06
SASKATCHEWAN	-0.11	-0.51	-0.41	0.12	-0.12	-0.08	-0.17
ALBERTA	0.21	-1.13	-0.34	-0.04	-0.71	-0.34	-0.43
BRITISH COLUMBIA	0.02	-1.11	-0.24	0.19	-0.47	-0.18	-0.19

Results by Province

decline of over 10 per cent in RDP. (Remember that in the case of a small province like Prince Edward Island, these results are more suspect.)

- RDP recovers to pre-1982 levels after 1986.

   . The most affected sectors are agriculture, fishing and construction.
   . Steady growth is projected during the rest of the 1980s and the 1990s. No immediate rebound is expected however.

- Employment grows moderately, about as fast as the labour force. As a result the unemployment rate changes little coming down from 13½ per cent to about 11½ per cent in 13 years.
- Per capita disposable income remains an almost constant 3/4 share of the national average throughout.

Nova Scotia experiences a severe recession with RDP falling 8¼ per cent in 1982, based on current employment figures. (Reconsideration of the output figures would suggest a less severe downturn in 1982 - and higher productivity - but perhaps less of a recovery in 1983, still leading as the present results do into the longer term.) However, development of Sable Island gas produces a fast recovery and strong employment growth.

- RDP

   . Strong growth is experienced in construction and utilities in the early years.
   . Mineral fuels production jumps in 1985-6 with Sable Island coming on stream.
   . Electrical products bounce back after a very poor 1982.
   . Chemical and petroleum products output grows very fast for the rest of the 1980s on the assumption of spinoffs from Sable Island (and Hibernia).

- Employment grows very fast in 1984 and 1985. The unemployment rate declines from 13½ per cent to just over 9 per cent by 1995 while remaining 2 percentage points above the national average.
- Real per capita disposable income rises from 80 to 85 per cent of the

The Ontario Economy 1982 - 1995

national average.

New Brunswick experiences a 4 1/3 per cent decline in RDP in 1982 from which it recovers steadily.

- RDP shows no dramatic winners or losers except perhaps the pulp and paper industry, which performs very well.
- Employment grows moderately so that the unemployment rate falls from 15 plus per cent to less than 11 per cent.
- Real disposable income remains 3/4 of the national average throughout.

Quebec experiences a very serious decline - over 9 per cent - in RDP in 1982, with a 14 per cent decline in the goods-producing sector. Again, this may be an overreaction to poor employment figures currently available. Some modest relative productivity gains could mean a smaller decline in output for the (known) decline in employment. As with Nova Scotia, probably both the extent of the 1982 decline and of the 1983 recovery are too large.

- Growth in manufacturing RDP remains weak, experiencing less than one per cent growth per annum in the 1990s.

    . Textiles and clothing in particular experience severe problems in the 1990s.
    . The food and beverage sector grows slowly.
    . Construction remains very weak.
    . Electric power and gas utilities grow quickly as the James Bay development continues and the gas pipeline extends natural gas use in the province.

- Employment grows quite slowly; however, the labour force grows even more slowly. As a result, the unemployment rate falls from over 14 per cent to over 9 per cent in 1995. Still, the unemployment rate remains well over 2 percentage points above the national average.
- Real per capita disposable income remains about 91 per cent of the national average.

Ontario manufacturing suffers a 7½ per cent decline in 1982 though a less

Results by Province

than 3 per cent decline in total RDP.

- . Sectors hard hit include wood and furniture, machinery and other transportation equipment, electrical products and non-metallic mineral products, all reflecting weak consumer and investment spending during the recession.
- . Construction never does recover strongly.
- . General goods-producing weakness is anticipated in the early to mid-1990s.

- Employment growth is strongest during the recovery period to 1985. The unemployment rate drops from 11 per cent to 6 per cent, approximately 1 percentage point below the national average.
- Real per capita disposable income ranges from 8½ per cent above the national average in 1982 to less than 1 per cent above by 1995. Average annual growth in real disposable income is the slowest in the country, but it is important to remember that, even so, the level of Ontario income remains above the national average.

Manitoba takes the recession hardest in the primary sector - particularly mining. The total decline in RDP in 1982 is expected to be just over 6 per cent. Recovery is relatively swift and strong.

- RDP falls most in the electrical products and non-metallic minerals sectors following a strong 1981. These sectors rebound quickly.

  - . Trade also contracts seriously in 1982 on top of a poor 1981.
  - . The government sector grows very slowly.
  - . The 1990s expose serious problems in specific manufacturing sectors, e.g., food and beverages; however, services look relatively good - as in the country as a whole.

- Employment and the labour force grow slowly except for a sharp upturn in employment in 1985. The unemployment rate falls from about 8 per cent to 5 3/4 per cent, well below the national average throughout.
- Real per capita disposable income remains essentially at 95 per cent of the national average over the outlook.

The Ontario Economy 1982 - 1995

Saskatchewan experiences a very fast recovery following one of the milder recessionary impacts in the country in 1982.

- RDP experiences a broadly based recovery. Even manufacturing does well relative to most provinces.
- Employment growth is very steady.
- Real per capita disposable income rises from 3½ per cent below to almost 9 per cent above the national average. Resources, energy, and agriculture all contribute to this result.

Alberta experiences only a 3 1/3 per cent setback in 1982, reflecting in part a 15 per cent decline in construction.

- . Manufacturing RDP also declines 5 1/3 per cent but bounces back quickly.
- . The important manufacturing sectors of food, primary metals and metal fabricating, and chemical and petrochemical products recover pre-recession levels during 1984 if not 1983.
- . Construction recovers by 1986, and then continues to grow. (A major danger to Alberta is a more severe setback to general construction than that projected here.)
- . Utilities provide an important source of growth through the 1990s.
- . Mineral fuels output varies around the 1981 level until 1993 when a major new synthetic facility comes on stream.

- Employment grows robustly after 1983. The unemployment rate falls from 8 per cent to a more usual 3 plus per cent.
- Real per capita disposable income climbs from 12 per cent to 22 per cent above the national average.

British Columbia's RDP tumbles 4 per cent in 1982; however, the recovery is very strong.

- . The pulp and paper industry recovers well.
- . Wood and furniture output recovers more slowly until North American construction picks up.
- . British Columbia construction recovers well from a 10 per cent

drop. Port development projects help.
- Electric power and gas utilities experience strong growth.
- The service sector appears very strong.
- Mineral fuels production jumps in 1984 and continues to grow in most years thereafter.

- Employment grows very fast. Unemployment drops from 12½ per cent to 7½ per cent.
- Real per capita income remains 10 per cent above the national average.

The Territories: The results for the Territories must be considered directionally indicative only. The results do suggest, however, that a very mild recession is experienced in the Territories.

- RDP

  - Construction output declines 7 plus per cent in 1982, but grows briskly thereafter.
  - Transportation and storage services also decline about 7 per cent in 1982, but recover quickly.
  - While difficult to measure exactly, Beaufort development in the 1980s and 1990s would appear to make a major impact on growth in the Territories.

7
Appendix – tables

INSTITUTE FOR POLICY ANALYSIS, UNIVERSITY OF TORONTO
FOCUS81E ** REVISED LO-TREND BASE CASE ** SEPT/82

Appendix Table 1
SUMMARY OF PROJECTION

(LEVELS)

	1981	1982	1983	1984	1985	1986	1987	1988
GROSS NATIONAL PRODUCT	328501	345982	386455	441871	512558	581140	653620	736414
IMPLICIT PRICE DEFLATOR FOR GNP (1971=1.0)	2.45	2.72	2.97	3.26	3.60	3.94	4.33	4.73
REAL GROSS NATIONAL PRODUCT	134070	127035	129947	135420	142321	147315	150815	155685
EXPENDITURE ON PERSONAL CONSUMPTION	83374	81173	81794	85279	89413	93142	96785	100311
EXPENDITURE BY GOVERNMENTS	26771	26939	27064	27595	27922	28319	28694	29153
INVESTMENT EXPENDITURE	27824	23970	23152	24378	26813	27377	27209	28078
RESIDENTIAL CONSTRUCTION	4997	4502	4445	5064	5786	5718	5696	5853
NON-RESIDENTIAL AND MACH. & EQUIP.	22827	19468	18707	19315	21027	21559	21513	22225
EXPORTS	32548	32620	33161	34220	35986	37431	38832	40048
IMPORTS	36733	34670	35347	36455	38487	39886	41674	42922
CAPACITY UTILIZATION RATE	0.79	0.70	0.75	0.74	0.76	0.77	0.76	0.76
UNEMPLOYMENT RATE	7.61	10.84	11.76	11.59	9.76	9.46	9.50	9.18
EMPLOYMENT (THOUSANDS)	10933	10584	10616	10872	11345	11631	11853	12043
NARROWLY DEFINED MONEY SUPPLY, M1	25067	26070	27373	29198	30918	33260	35640	37719
FINANCE CO. 90-DAY PAPER RATE	17.86	15.20	17.60	17.37	15.01	15.65	15.12	14.17
INDUSTRIAL BOND RATE	16.48	16.93	17.82	15.61	14.87	14.71	14.89	14.92
CONSUMER PRICE INDEX (1971=1.00)	2.37	2.63	2.86	3.14	3.46	3.81	4.21	4.60
AVERAGE ANNUAL WAGES AND SALARIES ($ '000)	14.799	16.231	17.683	19.201	20.939	22.938	25.248	27.737
GDP AT FACTOR COST PER EMPLOYEE ($71 '000)	10.948	10.747	11.007	11.193	11.255	11.341	11.378	11.553
EXCHANGE RATE (US $/CDN $)	0.834	0.808	0.791	0.794	0.802	0.802	0.805	0.812
TERMS OF TRADE (PX/PM) (1971=1.0)	1.04	1.02	1.03	1.02	1.04	1.03	1.04	1.04
BALANCE ON CURRENT ACCOUNT	-6576	-4472	-4538	-5755	-5156	-6536	-6776	-6393
CHANGE IN FOREIGN RESERVES	1426	-2000	0	0	0	0	0	0
CONSOLIDATED GOVERNMENT BALANCE	-2233	-20652	-25337	-25064	-19192	-17156	-17223	-15904
PERSONAL SAVINGS RATE (%)	11.52	13.18	13.95	13.01	11.90	10.55	9.57	9.34
NOMINAL AFTER-TAX CORPORATE PROFITS	21819	19543	28319	35381	43556	51954	56511	63382
REAL PERSONAL DISPOSABLE INCOME	97695	96667	97987	100784	104080	106611	109441	113011

NOTE - NOMINAL VALUES IN MILLIONS OF CURRENT DOLLARS
     - REAL VALUES IN MILLIONS OF 1971 DOLLARS

INSTITUTE FOR POLICY ANALYSIS, UNIVERSITY OF TORONTO
FOCUS81E ** REVISED LO-TREND BASE CASE ** SEPT/82

Appendix Table 1 (cont'd.)
SUMMARY OF PROJECTION

(LEVELS)

	1989	1990	1991	1992	1993	1994	1995
GROSS NATIONAL PRODUCT	821634	934021	1044016	1159062	1296133	1438895	1626595
IMPLICIT PRICE DEFLATOR FOR GNP (1971=1.0)	5.13	5.57	6.07	6.58	7.16	7.81	8.54
REAL GROSS NATIONAL PRODUCT	160256	167527	172043	176089	180920	184128	190430
EXPENDITURE ON PERSONAL CONSUMPTION	103254	106934	109610	111842	114876	117411	122182
EXPENDITURE BY GOVERNMENTS	29651	29989	30414	30857	31387	31870	32831
INVESTMENT EXPENDITURE	29266	32311	34218	35680	37190	38480	40583
RESIDENTIAL CONSTRUCTION	6046	6116	6095	6213	6383	6489	6396
NON-RESIDENTIAL AND MACH. & EQUIP.	23220	26196	28123	29467	30807	31992	34187
EXPORTS	41680	43684	45181	47306	49424	51220	52180
IMPORTS	44627	46641	48456	50514	52736	55552	57857
CAPACITY UTILIZATION RATE	0.76	0.77	0.77	0.77	0.77	0.77	0.78
UNEMPLOYMENT RATE	9.07	8.02	7.68	7.77	7.28	7.14	6.98
EMPLOYMENT (THOUSANDS)	12224	12619	12759	12927	13219	13435	13639
NARROWLY DEFINED MONEY SUPPLY, M1	39891	42063	44072	46576	49281	52493	55886
FINANCE CO. 90-DAY PAPER RATE	14.03	13.59	13.44	13.32	13.12	12.94	12.59
INDUSTRIAL BOND RATE	14.87	14.79	14.71	14.55	14.31	14.11	13.79
CONSUMER PRICE INDEX (1971=1.00)	5.02	5.50	6.03	6.60	7.21	7.88	8.57
AVERAGE ANNUAL WAGES AND SALARIES ($ '000)	30.351	33.203	36.410	39.853	43.656	47.947	52.807
GDP AT FACTOR COST PER EMPLOYEE ($71 '000)	11.725	11.882	12.054	12.168	12.212	12.210	12.405
EXCHANGE RATE (US $/CDN $)	0.810	0.805	0.800	0.795	0.797	0.810	0.825
TERMS OF TRADE (PX/PM) (1971=1.0)	1.05	1.04	1.04	1.04	1.05	1.07	1.10
BALANCE ON CURRENT ACCOUNT	-6915	-7165	-10012	-10193	-6239	-6194	-3848
CHANGE IN FOREIGN RESERVES	0	0	0	0	0	0	0
CONSOLIDATED GOVERNMENT BALANCE	-18276	-13618	-11305	-9333	-9222	-7980	-10818
PERSONAL SAVINGS RATE (%)	9.41	9.78	9.59	9.71	10.34	10.21	10.70
NOMINAL AFTER-TAX CORPORATE PROFITS	72514	87157	97815	105129	115558	125979	150069
REAL PERSONAL DISPOSABLE INCOME	116302	120801	123430	126021	130232	132820	138835

NOTE - NOMINAL VALUES IN MILLIONS OF CURRENT DOLLARS
    - REAL VALUES IN MILLIONS OF 1971 DOLLARS

INSTITUTE FOR POLICY ANALYSIS, UNIVERSITY OF TORONTO
FOCUS81E ** REVISED LO-TREND BASE CASE ** SEPT/82

Appendix Table 1 (cont'd.)
SUMMARY OF PROJECTION

(PERCENTAGE CHANGE; * INDICATES CHANGE IN LEVELS)

	1981	1982	1983	1984	1985	1986	1987	1988
GROSS NATIONAL PRODUCT	13.33	5.32	11.70	14.34	16.00	13.38	12.47	12.67
IMPLICIT PRICE DEFLATOR FOR GNP	10.05	11.09	9.25	9.67	10.39	9.55	9.86	9.15
REAL GROSS NATIONAL PRODUCT	3.00	-5.25	2.29	4.21	5.10	3.51	2.38	3.23
EXPENDITURE ON PERSONAL CONSUMPTION	1.73	-2.64	0.77	4.26	4.85	4.17	3.91	3.64
EXPENDITURE BY GOVERNMENTS	1.63	0.63	0.46	1.96	1.18	1.42	1.33	1.60
INVESTMENT EXPENDITURE	5.89	-13.85	-3.41	5.30	9.99	2.10	-0.62	3.20
RESIDENTIAL CONSTRUCTION	1.44	-9.91	-1.27	13.93	14.26	-1.17	-0.39	2.76
NON-RESIDENTIAL AND MACH. & EQUIP.	6.91	-14.72	-3.91	3.25	8.87	3.01	-0.68	3.31
EXPORTS	1.44	0.22	1.66	3.19	5.16	4.01	3.74	3.13
IMPORTS	3.14	-5.62	1.95	3.13	5.57	3.63	4.48	3.00
CAPACITY UTILIZATION RATE *	-0.02	-0.10	0.05	-0.00	0.02	0.00	-0.01	0.00
UNEMPLOYMENT RATE *	0.09	3.23	0.92	-0.17	-1.83	-0.30	0.04	-0.32
EMPLOYMENT	2.59	-3.19	0.30	2.41	4.36	2.52	1.91	1.60
NARROWLY DEFINED MONEY SUPPLY, M1	2.46	4.00	5.00	6.67	5.89	7.57	7.16	5.83
FINANCE CO. 90-DAY PAPER RATE *	4.01	-2.67	2.40	-0.22	-2.37	0.65	-0.53	-0.95
INDUSTRIAL BOND RATE *	3.13	0.46	0.88	-2.21	-0.74	-0.16	0.18	0.03
CONSUMER PRICE INDEX	12.45	11.18	8.46	10.10	9.96	10.30	10.29	9.31
AVERAGE ANNUAL WAGES AND SALARIES	9.88	9.68	8.95	8.58	9.05	9.55	10.07	9.86
PRODUCTIVITY CHANGE (GDP/EMPLOYEE)	1.40	-1.83	2.41	1.69	0.55	0.77	0.32	1.54
EXCHANGE RATE (US $/CDN $)	-2.50	-3.09	-2.06	0.38	0.96	0.00	0.30	0.89
TERMS OF TRADE (PX/PM)	-2.66	-1.96	0.57	-0.57	1.40	-0.66	1.07	0.38
BALANCE ON CURRENT ACCOUNT ($ MILL) *	-4672	2103	-65	-1217	599	-1379	-240	383
CHANGE IN FOREIGN RESERVES ($ MILL) *	2706	-3426	2000	0	0	0	0	0
CONSOLIDATED GOVERNMENT BALANCE ($ MILL) *	3750	-18419	-4684	273	5872	2036	-67	1319
PERSONAL SAVINGS RATE (%) *	1.41	1.66	0.77	-0.94	-1.11	-1.36	-0.98	-0.23
NOMINAL AFTER-TAX CORPORATE PROFITS	-14.65	-10.43	44.90	24.94	23.10	19.28	8.77	12.16
REAL PERSONAL DISPOSABLE INCOME	4.08	-1.05	1.37	2.85	3.27	2.43	2.65	3.26

INSTITUTE FOR POLICY ANALYSIS, UNIVERSITY OF TORONTO
FOCUS81E ** REVISED LO-TREND BASE CASE ** SEPT/82

Appendix Table 1 (cont'd.)
SUMMARY OF PROJECTION

(PERCENTAGE CHANGE; * INDICATES CHANGE IN LEVELS)

	1989	1990	1991	1992	1993	1994	1995
GROSS NATIONAL PRODUCT	11.57	13.68	11.78	11.02	11.83	11.01	13.04
IMPLICIT PRICE DEFLATOR FOR GNP	8.37	8.75	8.85	8.47	8.85	9.07	9.30
REAL GROSS NATIONAL PRODUCT	2.94	4.54	2.70	2.35	2.74	1.77	3.42
EXPENDITURE ON PERSONAL CONSUMPTION	2.93	3.56	2.50	2.04	2.71	2.21	4.06
EXPENDITURE BY GOVERNMENTS	1.71	1.14	1.42	1.45	1.72	1.54	3.02
INVESTMENT EXPENDITURE	4.23	10.41	5.90	4.27	4.23	3.47	5.46
RESIDENTIAL CONSTRUCTION	3.29	1.16	-0.34	1.94	2.73	1.66	-1.42
NON-RESIDENTIAL AND MACH. & EQUIP.	4.48	12.81	7.36	4.78	4.55	3.84	6.86
EXPORTS	4.07	4.81	3.43	4.70	4.48	3.63	1.87
IMPORTS	3.97	4.51	3.89	4.25	4.40	5.34	4.15
CAPACITY UTILIZATION RATE *	0.00	0.01	0.00	-0.00	-0.00	-0.00	0.01
UNEMPLOYMENT RATE *	-0.11	-1.05	-0.33	0.08	-0.48	-0.14	-0.16
EMPLOYMENT	1.50	3.23	1.11	1.31	2.26	1.64	1.51
NARROWLY DEFINED MONEY SUPPLY, M1	5.76	5.45	4.77	5.68	5.81	6.52	6.46
FINANCE CO. 90-DAY PAPER RATE *	-0.14	-0.44	-0.15	-0.13	-0.20	-0.18	-0.36
INDUSTRIAL BOND RATE *	-0.05	-0.09	-0.08	-0.16	-0.24	-0.21	-0.32
CONSUMER PRICE INDEX	9.18	9.45	9.76	9.40	9.32	9.27	8.76
AVERAGE ANNUAL WAGES AND SALARIES	9.43	9.40	9.66	9.46	9.54	9.83	10.14
PRODUCTIVITY CHANGE (GDP/EMPLOYEE)	1.48	1.34	1.45	0.95	0.37	-0.02	1.60
EXCHANGE RATE (US $/CDN $)	-0.24	-0.56	-0.64	-0.66	0.26	1.62	1.94
TERMS OF TRADE (PX/PM)	0.03	-0.04	-0.06	-0.15	1.01	1.82	2.75
BALANCE ON CURRENT ACCOUNT ($ MILL) *	-521	-250	-2847	-181	3955	44	2347
CHANGE IN FOREIGN RESERVES ($ MILL) *	0	0	0	0	0	0	0
CONSOLIDATED GOVERNMENT BALANCE ($ MILL) *	-2372	4658	2313	1972	111	1242	-2838
PERSONAL SAVINGS RATE (%) *	0.07	0.37	-0.20	0.12	0.63	-0.13	0.49
NOMINAL AFTER-TAX CORPORATE PROFITS	14.41	20.19	12.23	7.48	9.92	9.02	19.12
REAL PERSONAL DISPOSABLE INCOME	2.91	3.87	2.18	2.10	3.34	1.99	4.53

INSTITUTE FOR POLICY ANALYSIS, UNIVERSITY OF TORONTO
FOCUS81E ** REVISED LO-TREND BASE CASE ** SEPT/82

Appendix Table 2

GROSS NATIONAL PRODUCT IN CONSTANT DOLLARS (MILLIONS)

	1981	1982	1983	1984	1985	1986	1987	1988
PERSONAL CONSUMPTION EXPENDITURES. TOTAL	83374	81173	81794	85279	89413	93142	96785	100311
	(1.73)	(-2.64)	(0.77)	(4.26)	(4.85)	(4.17)	(3.91)	(3.64)
DURABLE GOODS	14347	12447	12607	13400	14539	15662	16903	17824
	(0.53)	(-13.25)	(1.29)	(6.29)	(8.50)	(7.73)	(7.93)	(5.45)
GOVERNMENT								
CURRENT	23227	23355	23464	23982	24276	24626	24955	25366
	(2.04)	(0.55)	(0.47)	(2.21)	(1.22)	(1.44)	(1.34)	(1.65)
GROSS CAPITAL FORMATION	3544	3584	3600	3612	3646	3693	3740	3787
	(-1.01)	(1.13)	(0.45)	(0.34)	(0.93)	(1.29)	(1.26)	(1.27)
BUSINESS GROSS FIXED CAPITAL FORMATION,TOTAL	27824	23970	23152	24378	26813	27377	27209	28078
	(5.89)	(-13.85)	(-3.41)	(5.30)	(9.99)	(2.10)	(-0.62)	(3.20)
RESIDENTIAL CONSTRUCTION	4997	4502	4445	5064	5786	5718	5696	5853
	(1.44)	(-9.91)	(-1.27)	(13.93)	(14.26)	(-1.17)	(-0.39)	(2.76)
NON-RESIDENTIAL CONSTRUCTION	10753	9207	8877	9396	10753	10697	10110	10471
	(8.43)	(-14.38)	(-3.58)	(5.84)	(14.45)	(-0.53)	(-5.49)	(3.57)
MACHINERY AND EQUIPMENT	12074	10261	9831	9919	10274	10963	11403	11755
	(5.60)	(-15.01)	(-4.20)	(0.90)	(3.58)	(6.70)	(4.02)	(3.08)
VALUE OF PHYSICAL CHANGE IN BUS. INVENTORIES	772	-2729	392	671	942	1200	1237	1284
	(1534)	(-3501)	(3122)	(279)	(272)	(258)	(38)	(47)
EXPORTS. TOTAL	32548	32620	33161	34220	35986	37431	38832	40048
	(1.44)	(0.22)	(1.66)	(3.19)	(5.16)	(4.01)	(3.74)	(3.13)
MERCHANDISE	27077	27320	27854	28653	30193	31370	32705	33890
	(4.08)	(0.90)	(1.96)	(2.87)	(5.37)	(3.90)	(4.25)	(3.62)
SERVICES	5471	5300	5306	5566	5794	6060	6127	6158
	(-9.90)	(-3.12)	(0.11)	(4.90)	(4.08)	(4.60)	(1.10)	(0.51)
IMPORTS, TOTAL	36733	34670	35347	36455	38487	39886	41674	42922
	(3.14)	(-5.62)	(1.95)	(3.13)	(5.57)	(3.63)	(4.48)	(3.00)
MERCHANDISE	23994	22182	22383	23311	24932	26341	28070	29060
	(1.36)	(-7.55)	(0.90)	(4.15)	(6.96)	(5.65)	(6.56)	(3.53)
SERVICES	12739	12488	12965	13145	13555	13544	13604	13862
	(6.67)	(-1.97)	(3.82)	(1.39)	(3.12)	(-0.08)	(0.44)	(1.90)
STATISTICAL DISCREPANCY	-486	-267	-268	-268	-268	-268	-268	-268
	(-362)	(218)	(0)	(0)	(0)	(0)	(0)	(0)
GROSS NATIONAL PRODUCT	134070	127035	129947	135420	142321	147315	150815	155685
	(3.00)	(-5.25)	(2.29)	(4.21)	(5.10)	(3.51)	(2.38)	(3.23)

NOTE - PERCENT CHANGES (OR LEVELS CHANGES) IN PARENTHESES

INSTITUTE FOR POLICY ANALYSIS, UNIVERSITY OF TORONTO
FOCUS81E ** REVISED LO-TREND BASE CASE ** SEPT/82

Appendix Table 2 (cont'd.)

GROSS NATIONAL PRODUCT IN CONSTANT DOLLARS (MILLIONS)

	1989	1990	1991	1992	1993	1994	1995
PERSONAL CONSUMPTION EXPENDITURES. TOTAL	103254	106934	109610	111842	114876	117411	122182
	(2.93)	(3.56)	(2.50)	(2.04)	(2.71)	(2.21)	(4.06)
DURABLE GOODS	18286	19048	19648	20128	20883	21571	22680
	(2.59)	(4.17)	(3.15)	(2.44)	(3.75)	(3.29)	(5.15)
GOVERNMENT							
CURRENT	25816	26107	26485	26882	27368	27807	28724
	(1.78)	(1.13)	(1.45)	(1.50)	(1.81)	(1.61)	(3.30)
GROSS CAPITAL FORMATION	3835	3882	3929	3975	4019	4063	4107
	(1.26)	(1.24)	(1.21)	(1.15)	(1.11)	(1.10)	(1.10)
BUSINESS GROSS FIXED CAPITAL FORMATION,TOTAL	29266	32311	34218	35680	37190	38480	40583
	(4.23)	(10.41)	(5.90)	(4.27)	(4.23)	(3.47)	(5.46)
RESIDENTIAL CONSTRUCTION	6046	6116	6095	6213	6383	6489	6396
	(3.29)	(1.16)	(-0.34)	(1.94)	(2.73)	(1.66)	(-1.42)
NON-RESIDENTIAL CONSTRUCTION	11213	13423	14578	15402	16030	16505	17914
	(7.09)	(19.70)	(8.61)	(5.65)	(4.08)	(2.96)	(8.54)
MACHINERY AND EQUIPMENT	12007	12773	13545	14065	14777	15487	16272
	(2.15)	(6.38)	(6.04)	(3.84)	(5.06)	(4.80)	(5.07)
VALUE OF PHYSICAL CHANGE IN BUS. INVENTORIES	1300	1518	1344	1188	1047	967	779
	(15)	(218)	(-174)	(-155)	(-140)	(-80)	(-188)
EXPORTS. TOTAL	41680	43684	45181	47306	49424	51220	52180
	(4.07)	(4.81)	(3.43)	(4.70)	(4.48)	(3.63)	(1.87)
MERCHANDISE	35420	37215	38633	40640	42585	44297	45246
	(4.51)	(5.07)	(3.81)	(5.19)	(4.79)	(4.02)	(2.14)
SERVICES	6261	6468	6548	6666	6839	6923	6934
	(1.66)	(3.32)	(1.23)	(1.80)	(2.60)	(1.23)	(0.15)
IMPORTS, TOTAL	44627	46641	48456	50514	52736	55552	57857
	(3.97)	(4.51)	(3.89)	(4.25)	(4.40)	(5.34)	(4.15)
MERCHANDISE	30385	32040	33680	35651	37867	40586	42985
	(4.56)	(5.44)	(5.12)	(5.85)	(6.22)	(7.18)	(5.91)
SERVICES	14242	14601	14775	14864	14869	14966	14872
	(2.74)	(2.52)	(1.19)	(0.60)	(0.03)	(0.66)	(-0.63)
STATISTICAL DISCREPANCY	-268	-268	-268	-268	-268	-268	-268
	(0)	(0)	(0)	(0)	(0)	(0)	(0)
GROSS NATIONAL PRODUCT	160256	167527	172043	176089	180920	184128	190430
	(2.94)	(4.54)	(2.70)	(2.35)	(2.74)	(1.77)	(3.42)

NOTE - PERCENT CHANGES (OR LEVELS CHANGES) IN PARENTHESES

INSTITUTE FOR POLICY ANALYSIS, UNIVERSITY OF TORONTO
FOCUS81E ** REVISED LO-TREND BASE CASE ** SEPT/82

Appendix Table 3

IMPLICIT PRICE INDEXES OF GROSS NATIONAL PRODUCT (1971=1.0)

	1981	1982	1983	1984	1985	1986	1987	1988
PERSONAL CONSUMPTION EXPENDITURES, TOTAL	2.28	2.54	2.74	3.03	3.31	3.62	3.97	4.32
	(11.13)	(11.33)	(7.99)	(10.49)	(9.24)	(9.52)	(9.51)	(8.82)
DURABLE GOODS	1.84	2.03	2.18	2.35	2.56	2.79	3.07	3.34
	(8.97)	(9.98)	(7.40)	(7.99)	(8.77)	(9.13)	(9.85)	(9.07)
GOVERNMENT								
CURRENT	2.85	3.21	3.55	3.85	4.23	4.62	5.06	5.53
	(12.00)	(12.76)	(10.35)	(8.65)	(9.73)	(9.34)	(9.59)	(9.18)
GROSS FIXED CAPITAL FORMATION	2.52	2.81	3.02	3.26	3.58	3.96	4.32	4.68
	(11.40)	(11.73)	(7.47)	(7.83)	(9.88)	(10.58)	(9.15)	(8.43)
BUSINESS GROSS FIXED CAPITAL FORMATION, TOTAL	2.56	2.88	3.13	3.46	3.89	4.34	4.78	5.25
	(11.19)	(12.51)	(8.63)	(10.68)	(12.42)	(11.45)	(10.26)	(9.71)
RESIDENTIAL CONSTRUCTION	3.23	3.72	4.09	4.49	5.07	5.71	6.33	7.00
	(14.76)	(15.12)	(10.01)	(9.95)	(12.73)	(12.71)	(10.92)	(10.60)
NON-RESIDENTIAL CONSTRUCTION	2.45	2.71	2.91	3.16	3.54	3.92	4.25	4.63
	(10.93)	(10.36)	(7.57)	(8.56)	(11.92)	(10.78)	(8.37)	(8.91)
MACHINERY AND EQUIPMENT	2.38	2.67	2.89	3.22	3.60	4.04	4.49	4.93
	(10.15)	(12.05)	(8.35)	(11.28)	(11.98)	(12.04)	(11.21)	(9.83)
EXPORTS, TOTAL	3.04	3.31	3.67	4.08	4.55	5.12	5.73	6.30
	(8.19)	(8.81)	(10.90)	(11.23)	(11.45)	(12.63)	(11.85)	(9.98)
MERCHANDISE	3.11	3.37	3.75	4.17	4.66	5.26	5.88	6.47
	(6.18)	(8.31)	(11.25)	(11.44)	(11.52)	(12.91)	(11.95)	(9.91)
SERVICES	2.71	3.02	3.28	3.63	4.01	4.46	4.93	5.41
	(16.69)	(11.39)	(8.75)	(10.67)	(10.52)	(11.09)	(10.68)	(9.73)
IMPORTS, TOTAL	2.92	3.24	3.57	3.99	4.39	4.98	5.51	6.03
	(11.18)	(10.93)	(10.32)	(11.84)	(9.93)	(13.36)	(10.66)	(9.57)
MERCHANDISE	3.21	3.55	3.92	4.40	4.84	5.49	6.03	6.62
	(11.21)	(10.64)	(10.32)	(12.20)	(10.06)	(13.33)	(9.91)	(9.85)
SERVICES	2.36	2.67	2.97	3.28	3.56	3.99	4.43	4.80
	(12.40)	(13.19)	(10.94)	(10.38)	(8.76)	(11.89)	(11.14)	(8.31)
GROSS NATIONAL PRODUCT	2.45	2.72	2.97	3.26	3.60	3.94	4.33	4.73
	(10.05)	(11.09)	(9.25)	(9.67)	(10.39)	(9.55)	(9.86)	(9.15)

NOTE - PERCENTAGE CHANGES ARE WRITTEN IN PARENTHESES

INSTITUTE FOR POLICY ANALYSIS, UNIVERSITY OF TORONTO
FOCUS81E ** REVISED LO-TREND BASE CASE ** SEPT/82

Appendix Table 3 (cont'd.)

IMPLICIT PRICE INDEXES OF GROSS NATIONAL PRODUCT (1971=1.0)

	1989	1990	1991	1992	1993	1994	1995
PERSONAL CONSUMPTION EXPENDITURES. TOTAL	4.67	5.06	5.50	5.96	6.47	7.01	7.57
	(8.15)	(8.40)	(8.67)	(8.32)	(8.52)	(8.40)	(7.97)
DURABLE GOODS	3.63	3.94	4.28	4.63	5.02	5.47	5.90
	(8.42)	(8.62)	(8.55)	(8.21)	(8.56)	(8.83)	(7.92)
GOVERNMENT							
CURRENT	6.01	6.55	7.12	7.75	8.42	9.17	9.90
	(8.61)	(9.04)	(8.75)	(8.80)	(8.68)	(8.93)	(7.94)
GROSS FIXED CAPITAL FORMATION	5.05	5.45	5.94	6.42	6.92	7.51	8.19
	(7.72)	(8.09)	(8.97)	(7.99)	(7.83)	(8.51)	(9.03)
BUSINESS GROSS FIXED CAPITAL FORMATION,TOTAL	5.71	6.21	6.80	7.37	7.96	8.65	9.44
	(-13.84)	(8.89)	(9.39)	(8.47)	(8.02)	(8.61)	(9.18)
RESIDENTIAL CONSTRUCTION	7.64	8.37	9.23	10.07	10.98	12.17	13.59
	(9.06)	(9.57)	(10.30)	(9.05)	(9.01)	(10.85)	(11.73)
NON-RESIDENTIAL CONSTRUCTION	4.99	5.47	5.99	6.43	6.82	7.22	7.85
	(7.94)	(9.62)	(9.39)	(7.31)	(6.21)	(5.76)	(8.71)
MACHINERY AND EQUIPMENT	5.40	5.95	6.57	7.22	7.90	8.70	9.57
	(9.45)	(10.31)	(10.41)	(9.90)	(9.44)	(10.11)	(10.01)
EXPORTS. TOTAL	6.88	7.54	8.23	8.94	9.70	10.51	11.39
	(9.17)	(9.51)	(9.17)	(8.66)	(8.54)	(8.31)	(8.35)
MERCHANDISE	7.06	7.73	8.42	9.13	9.90	10.70	11.58
	(9.08)	(9.51)	(9.01)	(8.46)	(8.40)	(8.08)	(8.16)
SERVICES	5.91	6.46	7.10	7.77	8.49	9.30	10.18
	(9.25)	(9.20)	(9.93)	(9.49)	(9.21)	(9.56)	(9.44)
IMPORTS, TOTAL	6.58	7.21	7.88	8.57	9.21	9.80	10.34
	(9.15)	(9.55)	(9.24)	(8.83)	(7.45)	(6.39)	(5.44)
MERCHANDISE	7.23	7.90	8.61	9.32	9.93	10.50	10.98
	(9.09)	(9.38)	(8.90)	(8.31)	(6.55)	(5.69)	(4.64)
SERVICES	5.22	5.70	6.23	6.79	7.39	7.92	8.46
	(8.78)	(9.30)	(9.18)	(9.02)	(8.81)	(7.24)	(6.82)
GROSS NATIONAL PRODUCT	5.13	5.57	6.07	6.58	7.16	7.81	8.54
	(8.37)	(8.75)	(8.85)	(8.47)	(8.85)	(9.07)	(9.30)

NOTE - PERCENTAGE CHANGES ARE WRITTEN IN PARENTHESES

INSTITUTE FOR POLICY ANALYSIS, UNIVERSITY OF TORONTO
FOCUS81E ** REVISED LO-TREND BASE CASE ** SEPT/82

Appendix Table 4

BALANCE OF INTERNATIONAL PAYMENTS AND EXCHANGE RATE

	1981	1982	1983	1984	1985	1986	1987	1988
	MILLIONS OF DOLLARS							
EXPORTS OF GOODS	84183	91994	104342	119616	140568	164910	192463	219201
	(10.52)	(9.28)	(13.42)	(14.64)	(17.52)	(17.32)	(16.71)	(13.89)
IMPORTS OF GOODS	77052	78813	87730	102514	120677	144498	169243	192469
	(12.71)	(2.29)	(11.31)	(16.85)	(17.72)	(19.74)	(17.12)	(13.72)
BALANCE OF TRADE	7131	13180	16612	17101	19891	20412	23221	26733
CAPITAL SERVICE RECEIPTS	1607	1826	2051	2287	2540	2868	3195	3499
	(-3.19)	(13.61)	(12.35)	(11.49)	(11.07)	(12.91)	(11.41)	(9.51)
CAPITAL SERVICE PAYMENTS	8589	10706	13387	14964	16443	17849	19472	21432
	(19.23)	(24.64)	(25.05)	(11.78)	(9.89)	(8.55)	(9.09)	(10.06)
BALANCE ON CAPITAL SERVICE	-6982	-8879	-11336	-12676	-13903	-14981	-16276	-17932
EXPORTS OF OTHER SERVICES	13209	14164	15356	17927	20710	24149	27033	29837
	(6.29)	(7.23)	(8.42)	(16.74)	(15.53)	(16.60)	(11.94)	(10.37)
IMPORTS OF OTHER SERVICES	21536	22716	25094	28100	31849	36136	40793	45078
	(20.46)	(5.48)	(10.47)	(11.98)	(13.34)	(13.46)	(12.89)	(10.50)
BALANCE ON SERVICES	-8327	-8551	-9738	-10173	-11138	-11986	-13759	-15241
NET TRANSFER RECEIPTS	1602	1588	1734	1803	1804	1830	1850	1858
CURRENT ACCOUNT BALANCE	-6576	-4472	-4538	-5755	-5156	-6536	-6776	-6393
NET LONG-TERM CAPITAL FLOWS	1340	-2372	-482	6038	4638	5062	4838	7520
NET SHORT-TERM CAPITAL FLOWS	6452	4845	5020	-282	518	1474	1938	-1126
SDRS AND ADJUSTMENT	210	0	0	0	0	0	0	0
CHANGE IN FOREIGN EXCHANGE RESERVES	1426	-2000	0	0	0	0	0	0
	DOLLARS							
EXCHANGE RATE: CDN$/US$	1.20	1.24	1.26	1.26	1.25	1.25	1.24	1.23
EXCHANGE RATE: WEIGHTED CDN$/WORLD	1.23	1.28	1.33	1.36	1.34	1.34	1.36	1.36
	1971=1.0							
TERMS OF TRADE (PX/PM)	1.04	1.02	1.03	1.02	1.04	1.03	1.04	1.04
TERMS OF TRADE -GOODS- (PXG/PMG)	0.97	0.98	0.98	0.97	0.98	0.97	0.99	0.99

NOTE - PERCENTAGE CHANGES ARE WRITTEN IN PARENTHESES

INSTITUTE FOR POLICY ANALYSIS, UNIVERSITY OF TORONTO
FOCUS81E ** REVISED LO-TREND BASE CASE ** SEPT/82

Appendix Table 4 (cont'd.)

BALANCE OF INTERNATIONAL PAYMENTS AND EXCHANGE RATE

	1989	1990	1991	1992	1993	1994	1995
MILLIONS OF DOLLARS							
EXPORTS OF GOODS	249905	287541	325383	371246	421669	474070	523743
	(14.01)	(15.06)	(13.16)	(14.10)	(13.58)	(12.43)	(10.48)
IMPORTS OF GOODS	219544	253203	289847	332294	376064	425998	472117
	(14.07)	(15.33)	(14.47)	(14.64)	(13.17)	(13.28)	(10.83)
BALANCE OF TRADE	30361	34338	35536	38952	45605	48072	51627
CAPITAL SERVICE RECEIPTS	3821	4150	4477	4824	5177	5522	5882
	(9.22)	(8.59)	(7.89)	(7.74)	(7.33)	(6.67)	(6.52)
CAPITAL SERVICE PAYMENTS	24269	27799	29794	32053	34170	35556	35436
	(13.24)	(14.55)	(7.18)	(7.58)	(6.61)	(4.05)	(-0.34)
BALANCE ON CAPITAL SERVICE	-20447	-23649	-25316	-27229	-28993	-30033	-29554
EXPORTS OF OTHER SERVICES	33203	37622	42011	46989	52878	58864	64687
	(11.28)	(13.31)	(11.67)	(11.85)	(12.53)	(11.32)	(9.89)
IMPORTS OF OTHER SERVICES	50068	55496	62230	68870	75668	83011	90420
	(11.07)	(10.84)	(12.13)	(10.67)	(9.87)	(9.70)	(8.92)
BALANCE ON SERVICES	-16864	-17874	-20219	-21881	-22789	-24147	-25733
NET TRANSFER RECEIPTS	1846	1831	1798	1775	1749	1724	1623
CURRENT ACCOUNT BALANCE	-6915	-7165	-10012	-10193	-6239	-6194	-3848
NET LONG-TERM CAPITAL FLOWS	7586	8218	6879	9244	8781	9438	8217
NET SHORT-TERM CAPITAL FLOWS	-671	-1052	3134	950	-2542	-3243	-4368
SDRS AND ADJUSTMENT	0	0	0	0	0	0	0
CHANGE IN FOREIGN EXCHANGE RESERVES	0	0	0	0	0	0	0
DOLLARS							
EXCHANGE RATE: CDN$/US$	1.23	1.24	1.25	1.26	1.26	1.23	1.21
EXCHANGE RATE: WEIGHTED CDN$/WORLD	1.38	1.40	1.42	1.44	1.44	1.41	1.39
1971=1.0							
TERMS OF TRADE (PX/PM)	1.05	1.04	1.04	1.04	1.05	1.07	1.10
TERMS OF TRADE -GOODS- (PXG/PMG)	0.98	0.97	0.96	0.95	0.95	0.96	0.98

NOTE - PERCENTAGE CHANGES ARE WRITTEN IN PARENTHESES

INSTITUTE FOR POLICY ANALYSIS, UNIVERSITY OF TORONTO
FOCUS81E   ** REVISED LO-TREND BASE CASE **   SEPT/82

Appendix Table 5
LABOUR MARKET VARIABLES

	1981	1982	1983	1984	1985	1986	1987	1988
THOUSANDS OF PERSONS								
POPULATION AGED 15 YEARS AND OVER	18295	18571	18816	19051	19291	19528	19737	19938
	(1.61)	(1.51)	(1.32)	(1.25)	(1.26)	(1.22)	(1.07)	(1.02)
LABOUR FORCE	11833	11872	12031	12297	12573	12846	13098	13260
	(2.69)	(0.32)	(1.35)	(2.21)	(2.24)	(2.18)	(1.95)	(1.24)
EMPLOYED	10933	10584	10616	10872	11345	11631	11853	12043
	(2.59)	(-3.19)	(0.30)	(2.41)	(4.36)	(2.52)	(1.91)	(1.60)
EMPLOYED IN PRIVATE SECTOR	9682	9316	9337	9570	10020	10281	10479	10644
	(2.80)	(-3.78)	(0.22)	(2.49)	(4.70)	(2.61)	(1.93)	(1.58)
GOVERNMENT CIVILIAN EMPLOYMENT	1250	1268	1279	1302	1326	1350	1374	1399
	(1.00)	(1.41)	(0.87)	(1.81)	(1.81)	(1.81)	(1.81)	(1.81)
CANADIAN ARMED FORCES	80	80	80	80	80	80	80	80
	(0.0)	(0.0)	(0.0)	(0.0)	(0.0)	(0.0)	(0.0)	(0.0)
THOUSANDS OF DOLLARS								
AVERAGE ANNUAL WAGE OF PRIVATE EMPLOYEES	14.799	16.231	17.683	19.201	20.939	22.938	25.248	27.737
	(9.88)	(9.68)	(8.95)	(8.58)	(9.05)	(9.55)	(10.07)	(9.86)
AVERAGE ANNUAL WAGE OF ARMED FORCES	23.200	25.421	27.868	30.528	33.495	36.847	40.608	44.463
	(11.61)	(9.57)	(9.62)	(9.55)	(9.72)	(10.01)	(10.21)	(9.49)
PER CENT								
UNEMPLOYMENT RATE	7.61	10.84	11.76	11.59	9.76	9.46	9.50	9.18
LABOUR FORCE PARTICIPATION RATES								
FEMALES 15-24 YEARS	63.27	61.59	62.69	60.60	61.62	62.30	62.97	63.17
FEMALES 25-54 YEARS	62.70	62.24	63.71	62.42	63.87	65.36	66.90	67.99
FEMALES 55 YEARS AND OVER	18.21	17.18	17.08	16.04	16.32	16.19	15.96	15.55
MALES 15-24 YEARS	72.49	72.24	74.85	73.13	73.90	74.61	75.28	75.43
MALES 25-54 YEARS	94.92	95.79	99.41	96.99	96.92	97.25	97.60	97.33
MALES 55 YEARS AND OVER	45.37	45.16	46.18	44.50	44.53	44.38	44.13	43.52
OVERALL RATE	64.68	63.92	63.94	64.54	65.17	65.79	66.36	66.51

NOTE -   PERCENTAGE CHANGES ARE WRITTEN IN PARENTHESES

INSTITUTE FOR POLICY ANALYSIS, UNIVERSITY OF TORONTO
FOCUS81E ** REVISED LO-TREND BASE CASE ** SEPT/82

Appendix Table 5 (cont'd.)

LABOUR MARKET VARIABLES

	1989	1990	1991	1992	1993	1994	1995
THOUSANDS OF PERSONS							
POPULATION AGED 15 YEARS AND OVER	20125	20318	20506	20697	20883	21066	21250
	(0.94)	(0.96)	(0.92)	(0.93)	(0.90)	(0.87)	(0.87)
LABOUR FORCE	13443	13719	13821	14015	14257	14468	14662
	(1.38)	(2.05)	(0.75)	(1.40)	(1.73)	(1.48)	(1.34)
EMPLOYED	12224	12619	12759	12927	13219	13435	13639
	(1.50)	(3.23)	(1.11)	(1.31)	(2.26)	(1.64)	(1.51)
EMPLOYED IN PRIVATE SECTOR	10800	11169	11283	11423	11688	11877	12052
	(1.46)	(3.42)	(1.02)	(1.25)	(2.32)	(1.62)	(1.48)
GOVERNMENT CIVILIAN EMPLOYMENT	1424	1450	1477	1503	1530	1558	1586
	(1.81)	(1.81)	(1.81)	(1.81)	(1.81)	(1.81)	(1.81)
CANADIAN ARMED FORCES	80	80	80	80	80	80	80
	(0.0)	(0.0)	(0.0)	(0.0)	(0.0)	(0.0)	(0.0)
THOUSANDS OF DOLLARS							
AVERAGE ANNUAL WAGE OF PRIVATE EMPLOYEES	30.351	33.203	36.410	39.853	43.656	47.947	52.807
	(9.43)	(9.40)	(9.66)	(9.46)	(9.54)	(9.83)	(10.14)
AVERAGE ANNUAL WAGE OF ARMED FORCES	48.482	52.928	57.839	63.113	68.792	74.883	81.236
	(9.04)	(9.17)	(9.28)	(9.12)	(9.00)	(8.85)	(8.48)
PER CENT							
UNEMPLOYMENT RATE	9.07	8.02	7.68	7.77	7.28	7.14	6.98
LABOUR FORCE PARTICIPATION RATES							
FEMALES 15-24 YEARS	63.48	64.71	64.61	65.07	66.04	66.70	67.30
FEMALES 25-54 YEARS	69.17	70.89	71.65	72.94	74.54	75.93	77.28
FEMALES 55 YEARS AND OVER	15.15	15.28	14.71	14.31	14.22	13.94	13.63
MALES 15-24 YEARS	75.67	76.43	76.26	76.66	77.35	77.86	78.34
MALES 25-54 YEARS	97.15	97.07	96.45	96.47	96.52	96.51	96.49
MALES 55 YEARS AND OVER	42.93	42.76	41.89	41.34	41.03	40.55	40.04
OVERALL RATE	66.80	67.52	67.40	67.72	68.27	68.68	69.00

NOTE - PERCENTAGE CHANGES ARE WRITTEN IN PARENTHESES

INSTITUTE FOR POLICY ANALYSIS, UNIVERSITY OF TORONTO
FOCUS81E ** REVISED LO-TREND BASE CASE ** SEPT/82

Appendix Table 6

FEDERAL GOVERNMENT REVENUES AND EXPENDITURES

	1981	1982	1983	1984	1985	1986	1987	1988
	MILLIONS OF DOLLARS							
REVENUES								
INDIRECT TAXES	18838	19937	22926	26461	29257	33744	40027	45870
	(55.29)	(5.83)	(14.99)	(15.42)	(10.57)	(15.34)	(18.62)	(14.60)
DIRECT TAXES - CORP. & GOV'T. BUS. ENT.	8624	5685	8229	9938	14532	18644	20736	23987
	(7.77)	(-34.08)	(44.74)	(20.78)	(46.23)	(28.29)	(11.22)	(15.68)
DIRECT TAXES & TRANSFERS FROM PERSONS	29088	33244	37829	42549	48233	54192	60110	66670
	(24.02)	(14.29)	(13.79)	(12.48)	(13.36)	(12.36)	(10.92)	(10.91)
DIRECT TAXES ON NONRESIDENTS	1110	949	1193	1277	1404	1570	1744	1934
INVESTMENT INCOME	5401	6313	5590	5504	6793	7085	7453	8009
NET TRANSF FROM (+), TO (-) OTHER GOV'TS *	-14060	-15568	-16767	-18070	-19530	-21099	-22791	-23444
	(9.60)	(10.73)	(7.71)	(7.77)	(8.08)	(8.03)	(8.02)	(2.86)
EXPENDITURES								
CURRENT EXP. ON GOODS & SERVICES	16300	18074	19848	22195	24721	27473	30588	34048
	(18.92)	(10.88)	(9.82)	(11.83)	(11.38)	(11.13)	(11.34)	(11.31)
GROSS CAPITAL FORMATION	1044	1325	1378	1486	1646	1844	2038	2237
	(2.96)	(26.89)	(4.04)	(7.79)	(10.82)	(11.98)	(10.52)	(9.80)
TRANSFERS TO PERSONS	18843	22953	26590	30171	33151	37609	42271	47375
	(13.69)	(21.81)	(15.85)	(13.47)	(9.88)	(13.45)	(12.40)	(12.07)
SUBSIDIES	6273	7430	9591	11466	12195	13976	16527	19614
	(14.35)	(18.45)	(29.08)	(19.55)	(6.36)	(14.60)	(18.26)	(18.67)
CAPITAL ASSISTANCE	761	1959	2856	3295	3622	4244	4967	5786
TRANSFERS TO NONRESIDENTS	857	966	1038	1131	1221	1319	1424	1538
INTEREST ON THE PUBLIC DEBT	13350	17137	19872	21083	22857	25176	27253	30126
	(38.30)	(28.37)	(15.96)	(6.09)	(8.42)	(10.14)	(8.25)	(10.54)
CAPITAL CONSUMPTION ALLOWANCES	923	990	1048	1342	1471	1626	1780	1947
SURPLUS (+) OR DEFICIT (-)	-7504	-18292	-21125	-21826	-17254	-15877	-16009	-15751
	(3193)	(-10788)	(-2833)	(-700)	(4572)	(1377)	(-132)	(259)

* INCLUDES 'OTHER REVENUES'

NOTE - PERCENT CHANGES (OR LEVELS CHANGES) IN PARENTHESES

INSTITUTE FOR POLICY ANALYSIS, UNIVERSITY OF TORONTO
FOCUS81E ** REVISED LO-TREND BASE CASE ** SEPT/82

Appendix Table 6 (cont'd.)

FEDERAL GOVERNMENT REVENUES AND EXPENDITURES

	1989	1990	1991	1992	1993	1994	1995
			MILLIONS OF DOLLARS				
REVENUES							
INDIRECT TAXES	52585	59842	71498	81680	94152	108134	124473
	(14.64)	(13.80)	(19.48)	(14.24)	(15.27)	(14.85)	(15.11)
DIRECT TAXES - CORP. & GOV'T. BUS. ENT.	27957	33626	38734	43084	47903	52896	55141
	(16.55)	(20.28)	(15.19)	(11.23)	(11.19)	(10.42)	(4.24)
DIRECT TAXES & TRANSFERS FROM PERSONS	73597	82548	91401	100627	110545	120497	135003
	(10.39)	(12.16)	(10.72)	(10.09)	(9.86)	(9.00)	(12.04)
DIRECT TAXES ON NONRESIDENTS	2129	2348	2574	2840	3133	3459	3742
INVESTMENT INCOME	8702	9369	10607	11807	13167	14752	16463
NET TRANSF FROM (+), TO (-) OTHER GOV'TS *	-24092	-24760	-25545	-27050	-28161	-29627	-31101
	(2.76)	(2.77)	(3.17)	(5.89)	(4.11)	(5.21)	(4.98)
EXPENDITURES							
CURRENT EXP. ON GOODS & SERVICES	37726	41664	46041	50933	56445	62553	70066
	(10.80)	(10.44)	(10.51)	(10.62)	(10.82)	(10.82)	(12.01)
GROSS CAPITAL FORMATION	2441	2671	2946	3218	3508	3848	4242
	(9.08)	(9.44)	(10.29)	(9.23)	(9.03)	(9.70)	(10.24)
TRANSFERS TO PERSONS	53228	59198	65757	72970	80708	89108	97944
	(12.35)	(11.22)	(11.08)	(10.97)	(10.60)	(10.41)	(9.92)
SUBSIDIES	23169	26403	33903	38916	45384	52737	63794
	(18.13)	(13.96)	(28.41)	(14.78)	(16.62)	(16.20)	(20.97)
CAPITAL ASSISTANCE	6726	7808	9094	10578	12312	14326	16599
TRANSFERS TO NONRESIDENTS	1661	1794	1938	2093	2260	2441	2636
INTEREST ON THE PUBLIC DEBT	33510	36956	41566	45927	50596	55685	61685
	(11.23)	(10.28)	(12.47)	(10.49)	(10.17)	(10.06)	(10.77)
CAPITAL CONSUMPTION ALLOWANCES	2121	2317	2547	2776	3017	3292	3616
SURPLUS (+) OR DEFICIT (-)	-15462	-11203	-9428	-8871	-7456	-7296	-9630
	(289)	(4258)	(1775)	(557)	(1415)	(160)	(-2333)

* INCLUDES 'OTHER REVENUES'

NOTE - PERCENT CHANGES (OR LEVELS CHANGES) IN PARENTHESES

INSTITUTE FOR POLICY ANALYSIS, UNIVERSITY OF TORONTO
FOCUS81E ** REVISED LO-TREND BASE CASE ** SEPT/82

Appendix Table 7

COMBINED PUBLIC SECTOR SURPLUS (+) OR DEFICIT (-)

	1981	1982	1983	1984	1985	1986	1987	1988
	MILLIONS OF DOLLARS							
FEDERAL	-7504	-18292	-21125	-21826	-17254	-15877	-16009	-15751
	(3193)	(-10788)	(-2833)	(-700)	(4572)	(1377)	(-132)	(259)
PROVINCIAL	2693	-4826	-6627	-5606	-4364	-3837	-4010	-3079
	(1128)	(-7519)	(-1801)	(1021)	(1242)	(527)	(-173)	(932)
LOCAL	-252	-458	-490	-428	-366	-280	-152	-83
	(-350)	(-206)	(-32)	(62)	(62)	(87)	(128)	(69)
PUBLIC HOSPITALS	-383	-13	-13	-13	-14	-14	-15	-15
	(-432)	(370)	(0)	(0)	(0)	(0)	(0)	(0)
CANADA & QUEBEC PENSION PLANS	3213	2937	2920	2810	2807	2853	2964	3025
	(211)	(-276)	(-17)	(-109)	(-3)	(46)	(111)	(60)
COMBINED PUBLIC SECTOR	-2233	-20652	-25337	-25064	-19192	-17156	-17223	-15904
	(3750)	(-18419)	(-4684)	(273)	(5872)	(2036)	(-67)	(1319)

NOTE - LEVELS CHANGES ARE WRITTEN IN PARENTHESES

INSTITUTE FOR POLICY ANALYSIS, UNIVERSITY OF TORONTO
FOCUS81E ** REVISED LO-TREND BASE CASE ** SEPT/82

Appendix Table 7

COMBINED PUBLIC SECTOR SURPLUS (+) OR DEFICIT (-)

	1989	1990	1991	1992	1993	1994	1995
MILLIONS OF DOLLARS							
FEDERAL	-15462	-11203	-9428	-8871	-7456	-7296	-9630
	(289)	(4258)	(1775)	(557)	(1415)	(160)	(-2333)
PROVINCIAL	-5791	-5388	-5019	-3808	-5442	-4656	-5587
	(-2712)	(402)	(370)	(1210)	(-1634)	(787)	(-931)
LOCAL	-61	-73	-143	-227	-417	-567	-882
	(21)	(-11)	(-69)	(-83)	(-190)	(-149)	(-314)
PUBLIC HOSPITALS	-15	-16	-16	-17	-17	-17	-18
	(0)	(0)	(0)	(0)	(0)	(0)	(0)
CANADA & QUEBEC PENSION PLANS	3054	3064	3302	3591	4112	4558	5300
	(30)	(10)	(238)	(289)	(521)	(446)	(741)
COMBINED PUBLIC SECTOR	-18276	-13618	-11305	-9333	-9222	-7980	-10818
	(-2372)	(4658)	(2313)	(1972)	(111)	(1242)	(-2838)

NOTE - LEVELS CHANGES ARE WRITTEN IN PARENTHESES

INSTITUTE FOR POLICY ANALYSIS, UNIVERSITY OF TORONTO
FOCUS81E ** REVISED LO-TREND BASE CASE ** SEPT/82

Appendix Table 8

NATIONAL AND PERSONAL INCOME

	1981	1982	1983	1984	1985	1986	1987	1988
MILLIONS OF CURRENT DOLLARS								
WAGES SALARIES AND SUPPLEMENTARY LABOUR INCOME	184752 (13.78)	198121 (7.24)	210922 (6.46)	237113 (12.42)	266005 (12.18)	299451 (12.57)	339572 (13.40)	379693 (11.82)
MILITARY PAY AND ALLOWANCES	1856 (11.61)	2033 (9.56)	2229 (9.63)	2442 (9.55)	2680 (9.72)	2948 (10.01)	3249 (10.21)	3557 (9.49)
CORPORATION PROFITS BEFORE TAXES AND BEFORE DIVIDENDS TO NON-RESIDENTS	33856 (-8.92)	27573 (-18.56)	40081 (45.36)	49442 (23.36)	63838 (29.12)	77882 (22.00)	85372 (9.62)	96627 (13.18)
NET INCOME OF UNINCORPORATED BUSINESS	17480 (13.45)	19097 (9.25)	26093 (36.63)	34429 (31.95)	43849 (27.36)	47778 (8.96)	51106 (6.97)	60622 (18.62)
OTHER INCOME	39412 (16.90)	44271 (12.33)	50179 (13.34)	61693 (22.95)	75171 (21.85)	82826 (10.18)	91130 (10.02)	106925 (17.33)
INVENTORY VALUATION ADJUSTMENT	-6721	-6225	-5766	-9650	-10206	-13798	-16618	-20689
NET NATIONAL INCOME	253220 (10.99)	265838 (4.98)	297709 (11.99)	341105 (14.58)	397552 (16.55)	449374 (13.04)	502770 (11.88)	566179 (12.61)
INDIRECT TAXES LESS SUBSIDIES	38241 (31.00)	39516 (3.33)	42834 (8.40)	48225 (12.59)	53819 (11.60)	60462 (12.34)	68679 (13.59)	76530 (11.43)
CAPITAL CONSUMPTION ALLOWANCES AND MISC. VALUATION ADJUSTMENTS	35766 (11.16)	39953 (11.71)	45242 (13.24)	51666 (14.20)	60225 (16.57)	70251 (16.65)	81008 (15.31)	92430 (14.10)
STATISTICAL DISCREPANCY	39512	40188	43501	49096	54777	61510	69836	77798
GROSS NATIONAL PRODUCT	328501 (13.33)	345982 (5.32)	386455 (11.70)	441871 (14.34)	512558 (16.00)	581140 (13.38)	653620 (12.47)	736414 (12.67)
PERSONAL INCOME	276631 (16.59)	307181 (11.04)	338210 (10.10)	383171 (13.29)	432947 (12.99)	485996 (12.25)	545014 (12.14)	611699 (12.24)
PERSONAL DISPOSABLE INCOME	222750 (15.67)	245394 (10.17)	268593 (9.45)	305290 (13.66)	344382 (12.80)	386300 (12.17)	434339 (12.44)	488005 (12.36)
PERSONAL SAVING	25750 (32.33)	32441 (25.98)	37494 (15.58)	39768 (6.07)	41012 (3.13)	40729 (-0.69)	41578 (2.09)	45575 (9.61)
PERSONAL SAVING RATE (PER CENT)	11.52 (13.92)	13.18 (14.43)	13.95 (5.85)	13.01 (-6.72)	11.90 (-8.54)	10.55 (-11.40)	9.57 (-9.27)	9.34 (-2.43)
PERSONAL DISPOSABLE INCOME IN CONSTANT (1971) DOLLARS	97695 (4.08)	96667 (-1.05)	97987 (1.37)	100784 (2.85)	104080 (3.27)	106611 (2.43)	109441 (2.65)	113011 (3.26)

NOTE - PERCENTAGE CHANGES ARE WRITTEN IN PARENTHESES

Appendix Table 8 (cont'd.)

INSTITUTE FOR POLICY ANALYSIS, UNIVERSITY OF TORONTO
FOCUS81E ** REVISED LO-TREND BASE CASE ** SEPT/82

NATIONAL AND PERSONAL INCOME

MILLIONS OF CURRENT DOLLARS

	1989	1990	1991	1992	1993	1994	1995
WAGES SALARIES AND SUPPLEMENTARY LABOUR INCOME	422228	471324 (11.20)	524495 (11.63)	583499 (11.68)	652903 (11.86)	733258 (12.31)	821835 (12.08)
MILITARY PAY AND ALLOWANCES	3879	4234 (9.17)	4627 (9.28)	5505 (9.12)	6049 (9.00)	6499 (8.85)	— (8.48)
CORPORATION PROFITS BEFORE TAXES AND BEFORE DIVIDENDS TO NON-RESIDENTS	111334 (15.22)	134082 (20.52)	165016 (13.29)	182257 (8.64)	199557 (10.35)	226467 (9.49)	— (13.48)
NET INCOME OF UNINCORPORATED BUSINESS	64723 (10.40)	77599 (9.35)	80789 (6.02)	91411 (10.93)	99962 (6.77)	111996 (−3.35)	119836 (24.04)
OTHER INCOME	116722 (9.16)	135229 (15.86)	165414 (9.12)	185961 (12.42)	196176 (5.49)	234630 (19.60)	—
INVENTORY VALUATION ADJUSTMENT	−24475	−35154	−45025	−48266	−35980	−63392	—
NET NATIONAL INCOME	629753 (11.23)	715091 (13.55)	795647 (11.27)	878149 (10.37)	978427 (11.42)	1081119 (10.50)	1226107 (13.41)
INDIRECT TAXES LESS SUBSIDIES	84983 (9.81)	95516 (12.39)	106430 (11.48)	118653 (11.48)	133966 (12.91)	150379 (16.75)	165138 (11.25)
CAPITAL CONSUMPTION ALLOWANCES AND MISC. VALUATION ADJUSTMENTS	105519 (13.52)	121914 (15.03)	140310 (13.29)	160490 (14.38)	181812 (12.92)	205301 (12.92)	233058
STATISTICAL DISCREPANCY	86356	97009	108051	120414	135885	152466	174193
GROSS NATIONAL PRODUCT	821634 (11.57)	930421 (13.68)	1044013 (11.82)	1159062 (13.04)	1296133	1438895	1625955
PERSONAL INCOME	678813 (10.97)	763658 (12.50)	848014 (11.05)	939947 (10.84)	1048013 (11.50)	1155687 (10.36)	1302850 (12.65)
PERSONAL DISPOSABLE INCOME	543224 (11.32)	611588 (12.58)	679020 (11.04)	751098 (10.61)	842134 (10.70)	931092 (10.36)	1050090 (12.87)
PERSONAL SAVING	51162	59853 (12.28)	65130 (18.28)	73036 (19.24)	87089 (12.14)	95052 (8.82)	112429 (12.26)
PERSONAL SAVING RATE (PER CENT)	9.41	9.78	9.59	9.71	10.34	10.21	10.70
PERSONAL DISPOSABLE INCOME IN CONSTANT (1971) DOLLARS	116799 (2.91)	120801 (3.87)	123820 (2.18)	126021 (2.10)	130232 (3.34)	132820 (1.99)	138835 (4.53)

NOTE – PERCENTAGE CHANGES ARE WRITTEN IN PARENTHESES

INSTITUTE FOR POLICY ANALYSIS, UNIVERSITY OF TORONTO
FOCUS81E ** REVISED LO-TREND BASE CASE ** SEPT/82

Appendix Table 9

CAPITAL FINANCE ACCOUNT

	1981	1982	1983	1984	1985	1986	1987	1988
	MILLIONS OF CURRENT DOLLARS							
GROSS DOMESTIC CAPITAL FORMATION								
BUSINESS GROSS FIXED CAPITAL FORMATION	71240	69047	72449	84437	104402	118805	130186	147398
	(17.73)	(-3.08)	(4.93)	(16.55)	(23.65)	(13.80)	(9.58)	(13.22)
CHANGE IN BUSINESS INVENTORIES	1565	-9300	881	1619	2769	4021	4772	5566
GOVERNMENT GROSS CAPITAL FORMATION	8931	10075	10875	11768	13052	14618	16157	17741
	(10.48)	(12.81)	(7.94)	(8.21)	(10.92)	(12.00)	(10.52)	(9.80)
FINANCING OF CAPITAL FORMATION								
PERSONAL SAVING	25750	32441	37494	39768	41012	40729	41578	45575
	(32.33)	(25.98)	(15.58)	(6.07)	(3.13)	(-0.69)	(2.09)	(9.61)
UNDISTRIBUTED CORPORATE PROFITS	13764	12684	19875	22260	27980	33043	35057	39607
	(-26.81)	(-7.84)	(56.69)	(12.00)	(25.70)	(18.09)	(6.10)	(12.98)
PUBLIC SECTOR SURPLUS (+) OR DEFICIT (-)	-2233	-20652	-25337	-25064	-19192	-17156	-17223	-15904
CURRENT ACCOUNT DEFICIT (+) OR SURPLUS (-) WITH NON-RESIDENTS	6576	4473	4538	5756	5157	6536	6776	6394
CAPITAL CONSUMPTION ALLOWANCES (NON-GOV'T)	30366	33670	38146	43947	51763	60898	70772	81230
	(10.22)	(10.88)	(13.29)	(15.21)	(17.78)	(17.65)	(16.21)	(14.78)
INVENTORY VALUATION ADJUSTMENT	-6721	-6225	-5766	-9650	-10206	-13798	-16618	-20689
OTHER	14234	13432	15255	20807	23710	27193	30773	34493
	(14.43)	(-5.63)	(13.57)	(36.39)	(13.95)	(14.69)	(13.16)	(12.09)

NOTE - PERCENTAGE CHANGES ARE WRITTEN IN PARENTHESES

INSTITUTE FOR POLICY ANALYSIS, UNIVERSITY OF TORONTO
FOCUS81E ** REVISED LO-TREND BASE CASE ** SEPT/82

Appendix Table 9 (cont'd.)

CAPITAL FINANCE ACCOUNT

	1989	1990	1991	1992	1993	1994	1995
MILLIONS OF CURRENT DOLLARS							
GROSS DOMESTIC CAPITAL FORMATION							
BUSINESS GROSS FIXED CAPITAL FORMATION	167014	200785	232598	263078	296209	332859	383263
	(13.31)	(20.22)	(15.84)	(13.10)	(12.59)	(12.37)	(15.14)
CHANGE IN BUSINESS INVENTORIES	6317	8438	8082	7617	6971	5572	6444
GOVERNMENT GROSS CAPITAL FORMATION	19352	21178	23357	25513	27816	30513	33637
	(9.08)	(9.44)	(10.29)	(9.23)	(9.03)	(9.70)	(10.24)
FINANCING OF CAPITAL FORMATION							
PERSONAL SAVING	51162	59853	65130	73036	87089	95052	112429
	(12.26)	(16.99)	(8.82)	(12.14)	(19.24)	(9.14)	(18.28)
UNDISTRIBUTED CORPORATE PROFITS	45693	55744	62231	66286	72643	78941	89847
	(15.37)	(22.00)	(11.64)	(6.52)	(9.59)	(8.67)	(13.82)
PUBLIC SECTOR SURPLUS (+) OR DEFICIT (-)	-18276	-13618	-11305	-9332	-9222	-7980	-10818
CURRENT ACCOUNT DEFICIT (+) OR SURPLUS (-) WITH NON-RESIDENTS	6915	7165	10012	10194	6239	6195	3848
CAPITAL CONSUMPTION ALLOWANCES (NON-GOV'T)	93323	108585	125661	144523	164460	186367	212259
	(14.89)	(16.35)	(15.73)	(15.01)	(13.80)	(13.32)	(13.89)
INVENTORY VALUATION ADJUSTMENT	-24475	-29945	-35154	-41035	-48266	-53930	-63392
OTHER	38343	42616	47462	52538	58054	64300	79169
	(11.16)	(11.15)	(11.37)	(10.69)	(10.50)	(10.76)	(23.12)

NOTE - PERCENTAGE CHANGES ARE WRITTEN IN PARENTHESES

INSTITUTE FOR POLICY ANALYSIS, UNIVERSITY OF TORONTO
ENR82A   ***REVISED LO-TREND BASE CASE *** SEPT/82

Appendix Table 10
ENERGY MODULE - SUMMARY TABLE

PRODUCTION/DEMAND
(MILL. BBL. OR BILL. CU.FT.)

	1982	1983	1984	1985	1986	1987	1988
PETROLEUM PRODUCTION							
CONVENTIONAL CRUDE (OLD)	426	410	386	365	343	322	302
CONVENTIONAL CRUDE (NEW)	34	41	47	53	57	67	78
SYNTHETIC	55	55	65	65	65	65	65
TERTIARY	0	0	0	0	0	1	1
FRONTIER-HIBERNIA	0	0	0	0	0	32	63
TOTAL PETROLEUM PRODUCTION	515	506	498	482	466	487	509
DOMESTIC DEMAND FOR PETROLEUM	604	599	592	597	597	593	587
TOTAL NATURAL-GAS PRODUCTION	2712	2967	3263	3418	3595	3697	3747
DOMESTIC DEMAND FOR NATURAL GAS	1912	1967	2063	2168	2295	2347	2397

DOMESTIC PRICES
(CDN.$/BBL. OR CDN.$/THOUS. CU. FT.)

	1982	1983	1984	1985	1986	1987	1988
WELLHEAD PRICE, CONVENTIONAL CRUDE (OLD)	24.63	31.02	37.50	44.15	52.25	60.93	70.47
ENERGY AGREEMENT BASE FOR WELLHEAD PRICE	24.63	31.75	39.75	47.75	55.75	64.50	76.00
REFINERS' ACQUISITION PRICE (BLENDED PRICE)	31.48	35.59	42.79	50.01	59.76	70.35	82.04
NEW OIL REFERENCE PRICE	47.57	55.06	61.83	68.78	75.78	71.65	79.68
OIL IMPORT PRICE	42.07	44.22	51.31	60.22	71.07	82.69	95.47
PRICE FOR NEW CONVENTIONAL CRUDE	42.07	44.22	51.31	60.22	71.07	82.69	95.47
PRICE FOR TERTIARY PRODUCTION	42.07	44.22	51.31	60.22	71.07	82.69	95.47
PRICE FOR FRONTIER-HIBERNIA PRODUCTION	42.07	44.22	51.31	60.22	71.07	82.69	95.47
PRICE FOR SYNTHETIC-OIL PRODUCTION	42.07	44.22	51.31	60.22	71.07	82.69	95.47
NETBACK ON NATURAL GAS, DOMESTIC SALES	2.19	2.69	3.19	3.69	4.19	5.08	6.00
NETBACK ON NATURAL GAS, EXPORT SALES	5.14	5.42	6.43	7.77	9.40	11.16	13.10
WEIGHTED NETBACK PRICE - NATURAL GAS	3.07	3.62	4.38	5.18	6.08	7.30	8.56
TORONTO CITYGATE PRICE, NATURAL GAS	3.58	4.04	4.80	5.61	6.70	7.88	9.19

Appendix Table 10 (Cont'd.)

INSTITUTE FOR POLICY ANALYSIS, UNIVERSITY OF TORONTO
ENR82A ***REVISED LO-TREND BASE CASE * SEPT/82**

ENERGY MODULE - SUMMARY TABLE

PRODUCTION/DEMAND
(MILL. BBL. OR BILL. CU.FT.)

	1989	1990	1991	1992	1993	1994	1995
PETROLEUM PRODUCTION							
CONVENTIONAL CRUDE (OLD)	286	271	267	261	259	249	249
CONVENTIONAL CRUDE (NEW)	84	87	89	92	94	96	101
SYNTHETIC	65	65	65	65	105	105	115
TERTIARY	4	7	8	14	28	41	53
FRONTIER-HIBERNIA	75	75	75	75	75	75	94
TOTAL PETROLEUM PRODUCTION	514	505	505	507	560	566	612
DOMESTIC DEMAND FOR PETROLEUM	584	582	591	597	605	606	616
TOTAL NATURAL-GAS PRODUCTION	3799	3853	3969	4062	4166	4225	4329
DOMESTIC DEMAND FOR NATURAL GAS	2449	2503	2619	2712	2816	2875	2979

DOMESTIC PRICES
(CDN.$/BBL. OR CDN.$/THOUS. CU. FT.)

	1989	1990	1991	1992	1993	1994	1995
WELLHEAD PRICE, CONVENTIONAL CRUDE (OLD)	81.67	94.38	108.35	123.11	136.00	148.00	160.00
ENERGY AGREEMENT BASE FOR WELLHEAD PRICE	88.00	100.00	112.00	124.00	136.00	148.00	160.00
REFINERS' ACQUISITION PRICE (BLENDED PRICE)	95.71	111.29	128.27	147.11	165.96	184.63	206.57
NEW OIL REFERENCE PRICE	89.04	99.82	105.92	118.79	132.75	146.87	162.84
OIL IMPORT PRICE	110.44	127.44	146.12	167.43	190.04	212.13	235.68
PRICE FOR NEW CONVENTIONAL CRUDE	110.44	127.44	146.12	167.43	190.04	212.13	235.68
PRICE FOR TERTIARY PRODUCTION	110.44	127.44	146.12	167.43	190.04	212.13	235.68
PRICE FOR FRONTIER-HIBERNIA PRODUCTION	110.44	127.44	146.12	167.43	190.04	212.13	235.68
PRICE FOR SYNTHETIC-OIL PRODUCTION	110.44	127.44	146.12	167.43	190.04	212.13	235.68
NETBACK ON NATURAL GAS, DOMESTIC SALES	6.85	7.85	8.92	10.33	11.92	13.67	16.20
NETBACK ON NATURAL GAS, EXPORT SALES	17.37	20.80	24.05	27.51	30.90	34.53
WEIGHTED NETBACK PRICE - NATURAL GAS	9.88	11.39	12.96	14.89	16.97	19.18	21.32
TORONTO CITYGATE PRICE, NATURAL GAS	10.73	12.47	14.38	16.49	18.60	20.69	23.15

INSTITUTE FOR POLICY ANALYSIS, UNIVERSITY OF TORONTO
FOCUS81E ** REVISED LO-TREND BASE CASE ** SEPT/82

Appendix Table 11
ENERGY IMPACTS - SUMMARY TABLE

EXTERNAL TRADE

	1981	1982	1983	1984	1985	1986	1987	1988
EXCHANGE RATE (CDN$/US$)	1.20	1.24	1.26	1.26	1.25	1.25	1.24	1.23
	(2.57)	(3.18)	(2.11)	(-0.38)	(-0.95)	(-0.00)	(-0.30)	(-0.88)

($71 MILLION)

EXPORTS OF OIL & GAS	421	425	488	549	502	531	558	557
	(-6.53)	(0.95)	(14.95)	(12.42)	(-8.53)	(5.69)	(5.15)	(-0.12)
IMPORTS OF OIL	422	328	349	381	349	394	339	275
	(-0.03)	(-22.40)	(6.40)	(9.40)	(-8.40)	(12.68)	(-13.77)	(-19.08)

($ MILLIONS)

EXPORTS OF OIL & GAS	7234	7844	9398	12075	12489	15488	18851	22008
	(5.23)	(8.44)	(19.81)	(28.49)	(3.43)	(24.01)	(21.72)	(16.75)
IMPORTS OF OIL	8119	6263	6851	8540	9035	11821	11691	10805
	(16.61)	(-22.86)	(9.40)	(24.65)	(5.79)	(30.84)	(-1.10)	(-7.57)
BALANCE ON OIL AND GAS TRADE	-885	1581	2546	3535	3454	3667	7160	11203

(1971=100.(CDN))

DEFLATOR; OIL & GAS EXPORTS ($US)	1438	1492	1524	1747	1995	2341	2719	3206
	(8.82)	(3.77)	(2.16)	(14.62)	(14.21)	(17.34)	(16.14)	(17.94)
DEFLATOR; OIL & GAS EXPORTS ($CDN)	1723	1846	1925	2199	2488	2919	3379	3950
	(11.57)	(7.14)	(4.28)	(14.21)	(13.13)	(17.33)	(15.77)	(16.89)
DEFLATOR; OIL IMPORTS ($US)	1603	1544	1556	1777	2070	2408	2776	3198
	(13.40)	(-3.70)	(0.75)	(14.20)	(16.50)	(16.33)	(15.29)	(15.20)
DEFLATOR; OIL IMPORTS ($CDN)	1922	1911	1966	2237	2581	3002	3450	3940
	(16.24)	(-0.59)	(2.87)	(13.79)	(15.40)	(16.32)	(14.92)	(14.18)

DOMESTIC PRICES

	1981	1982	1983	1984	1985	1986	1987	1988
REFINERS' ACQUISITION PRICE ($/BBL)	26.54	31.94	36.07	42.79	50.01	59.76	70.35	82.04
	(55.87)	(20.35)	(12.92)	(18.64)	(16.88)	(19.50)	(17.71)	(16.62)
CPI - GASOLINE (1971=1.0)	3.27	4.23	4.99	5.66	6.23	7.00	7.83	8.76
	(36.03)	(29.48)	(18.04)	(13.39)	(10.07)	(12.35)	(11.93)	(11.78)

NOTE - PERCENTAGE CHANGES ARE WRITTEN IN PARENTHESES

INSTITUTE FOR POLICY ANALYSIS, UNIVERSITY OF TORONTO
FOCUS81E ** REVISED LO-TREND BASE CASE ** SEPT/82

Appendix Table 11 (cont'd.)
ENERGY IMPACTS - SUMMARY TABLE

NOTE - PERCENTAGE CHANGES ARE WRITTEN IN PARENTHESES

EXTERNAL TRADE

	1989	1990	1991	1992	1993	1994	1995
EXCHANGE RATE (CDN$/US$)	1.23	1.24	1.25	1.26	1.26	1.23	1.21
	(0.24)	(0.57)	(0.64)	(0.66)	(-0.26)	(-1.59)	(-1.90)

EXPORTS OF OIL & GAS ($ MILLION)

	1989	1990	1991	1992	1993	1994	1995
EXPORTS OF OIL & GAS	560	563	564	566	569	571	618
	(0.48)	(0.64)	(0.16)	(0.38)	(0.52)	(0.32)	(8.24)
IMPORTS OF OIL	256	275	302	314	202	190	151
	(-6.98)	(7.53)	(9.94)	(3.85)	(-35.57)	(-6.08)	(-20.46)

EXPORTS OF OIL & GAS ($ MILLIONS)

EXPORTS OF OIL & GAS	25692	29868	34459	39696	45267	50730	61506
	(16.73)	(16.26)	(15.37)	(15.20)	(14.03)	(12.07)	(21.24)
IMPORTS OF OIL	14534	17201	17763	20983	15246	15834	13883
	(6.74)	(23.13)	(25.08)	(18.13)	(-27.34)	(4.15)	(-12.57)
BALANCE ON OIL AND GAS TRADE	14158	15667	16697	18713	30021	34852	47624

DEFLATOR: OIL & GAS EXPORTS (US$) (1971=100.(CDN))

	3716	4268	4885	5570	6334	7191	8210
	(15.89)	(14.73)	(14.45)	(14.10)	(13.72)	(13.53)	(14.17)

DEFLATOR: OIL & GAS EXPORTS ($CDN)

	4589	5301	6106	7008	7950	8881	9947
	(16.17)	(15.52)	(15.19)	(14.76)	(13.44)	(11.72)	(11.99)

DEFLATOR: OIL IMPORTS (US$)

	3655	4158	4701	5316	6013	6785	7649
	(14.29)	(13.76)	(13.07)	(13.07)	(13.13)	(12.84)	(12.73)

DEFLATOR: OIL IMPORTS ($CDN)

	4514	5164	5876	6698	8899	8380	9267
	(14.57)	(14.41)	(13.80)	(13.82)	(12.83)	(11.04)	(10.58)

DOMESTIC PRICES

REFINERS' ACQUISITION PRICE ($/BBL)

	1989	1990	1991	1992	1993	1994	1995
	95.71	111.29	128.27	147.11	165.96	184.63	206.57
	(16.66)	(16.28)	(15.26)	(14.69)	(12.81)	(11.25)	(11.88)

CPI - GASOLINE (1971=1.0)

	9.84	11.07	12.41	13.89	15.38	16.85	18.59
	(12.32)	(12.50)	(11.98)	(11.91)	(10.70)	(9.58)	(10.27)

INSTITUTE FOR POLICY ANALYSIS, UNIVERSITY OF TORONTO
PRISM82A   *** REVISED LO-TREND BASE CASE ***   SEPT/82

Appendix Table 12
CANADA: REAL DOMESTIC PRODUCT BY INDUSTRY

(MILLIONS OF 1971 DOLLARS)

	1981	1982	1983	1984	1985	1986	1987	1988
1 AGRICULTURE, FISHING & TRAPPING	3319	3178	3252	3268	3322	3348	3375	3434
2 FORESTRY	735	675	686	691	724	737	745	762
3 MINERAL FUEL MINES & WELLS	1416	1461	1508	1511	1519	1561	1604	1623
4 OTHER MINES & QUARRIES	1833	1662	1665	1646	1701	1703	1694	1715
5 FOOD, FEED, BEVERAGES & TOBACCO	3550	3401	3490	3558	3615	3643	3666	3732
6 TEXTILES & CLOTHING	1799	1668	1704	1788	1876	1940	1951	1999
7 WOOD & FURNITURE	1605	1451	1469	1482	1563	1588	1599	1631
8 PAPER, PRINTING & ALLIED INDUSTRIES	3600	3413	3496	3522	3656	3744	3816	3920
9 PRIMARY METAL & METAL FABRICATING	4347	3943	3983	4059	4254	4320	4311	4407
10 MOTOR VEHICLES & PARTS	2060	1903	1875	2251	2411	2473	2421	2430
11 MACHINERY & OTH. TRANSPORTATION EQUIP.	2581	2241	2257	2248	2311	2415	2463	2552
12 ELECTRICAL PRODUCTS	1784	1498	1505	1530	1588	1629	1632	1662
13 CHEMICAL, RUBBER & PETROLEUM PRODUCTS	2959	2818	2929	3070	3232	3342	3407	3547
14 NON-METALLIC MINERAL PRODUCTS	987	880	890	933	1008	1014	996	1024
15 OTHER MANUFACTURING INDUSTRIES	1008	901	921	940	968	990	990	1009
TOTAL MANUFACTURING	26279	24117	24519	25380	26482	27099	27252	27914
16 CONSTRUCTION	7203	6275	6295	6639	7284	7241	7028	7171
17 ELECTRIC POWER, WATER & GAS UTILITIES	3915	3791	3988	4213	4449	4646	4807	5021
TOTAL GOODS-PRODUCING INDUSTRIES	44699	41159	41913	43349	45481	46335	46506	47640
18 TRANSPORTATION & STORAGE	7766	7221	7415	7771	8116	8410	8533	8701
19 COMMUNICATION	5384	5294	5649	6102	6626	7136	7606	8140
20 TRADE	15010	13799	14239	14866	15697	16336	16850	17461
21 FINANCE, INSURANCE & REAL ESTATE	15884	15503	16067	16863	17821	18603	19286	20011
22 OTHER SERVICE INDUSTRIES	17240	16828	17408	18358	19374	20308	21085	21953
TOTAL SERVICE INDUSTRIES	61284	58645	60778	63959	67634	70793	73361	76266
23 GOVERNMENT SECTOR	13710	13951	14159	14379	14576	14781	14998	15233
TOTAL: ALL SECTORS (GROSS DOMESTIC PRODUCT AT FACTOR COST - REAL)	119693	113754	116850	121687	127691	131909	134864	139140

INSTITUTE FOR POLICY ANALYSIS, UNIVERSITY OF TORONTO
PRISM82A * REVISED LO-TREND BASE CASE *** SEPT/82**

Appendix Table 12 (cont'd.)
CANADA: REAL DOMESTIC PRODUCT BY INDUSTRY

(YEAR-OVER-YEAR PER-CENT CHANGE)

	1989	1990	1991	1992	1993	1994	1995
1 AGRICULTURE,FISHING & TRAPPING	1.61	2.31	0.08	0.53	1.42	-0.15	0.19
2 FORESTRY	2.60	4.66	1.31	1.85	2.14	0.89	1.02
3 MINERAL FUEL MINES & WELLS	0.0	1.73	1.51	5.91	1.30	4.95	6.36
4 OTHER MINES & QUARRIES	1.58	3.72	0.32	0.95	0.88	-0.57	-0.80
5 FOOD,FEED,BEVERAGES & TOBACCO	1.47	2.15	0.21	0.11	1.38	-0.23	0.87
6 TEXTILES & CLOTHING	1.60	2.27	0.25	-0.78	-0.51	-1.52	-0.43
7 WOOD & FURNITURE	2.47	5.07	1.48	1.86	2.17	1.09	1.26
8 PAPER,PRINTING & ALLIED INDUSTRIES	2.84	3.90	1.30	2.01	2.29	1.03	0.89
9 PRIMARY METAL & METAL FABRICATING	2.01	4.92	1.64	1.28	1.07	-0.64	-0.11
10 MOTOR VEHICLES & PARTS	-0.09	3.93	5.09	0.13	1.27	0.04	-1.78
11 MACHINERY & OTH. TRANSPORTATION EQUIP.	1.52	4.62	2.23	1.29	0.87	-1.59	-0.42
12 ELECTRICAL PRODUCTS	0.18	3.09	0.69	-0.53	-0.25	-1.68	0.22
13 CHEMICAL,RUBBER & PETROLEUM PRODUCTS	3.47	5.07	2.50	2.05	2.48	0.50	1.90
14 NON-METALLIC MINERAL PRODUCTS	3.24	7.65	2.79	1.86	1.34	-0.21	1.87
15 OTHER MANUFACTURING INDUSTRIES	-0.13	0.80	-1.60	-2.35	-2.42	-4.35	-1.56
TOTAL MANUFACTURING	1.87	3.98	1.64	0.92	1.25	-0.33	0.38
16 CONSTRUCTION	3.64	9.26	3.53	2.44	2.01	1.10	3.44
17 ELECTRIC POWER,WATER & GAS UTILITIES	4.39	5.63	3.62	3.39	4.45	3.10	4.28
TOTAL GOODS-PRODUCING INDUSTRIES	2.32	4.77	1.99	1.59	1.75	0.50	1.51
18 TRANSPORTATION & STORAGE	2.23	4.11	1.69	1.63	2.29	0.88	1.50
19 COMMUNICATION	6.73	8.09	6.18	5.83	6.19	5.14	7.04
20 TRADE	3.07	4.53	2.55	2.06	3.07	2.25	3.25
21 FINANCE,INSURANCE & REAL ESTATE	3.43	4.55	2.60	2.29	2.52	1.48	3.86
22 OTHER SERVICE INDUSTRIES	3.96	5.31	3.43	3.10	3.56	2.37	5.54
TOTAL SERVICE INDUSTRIES	3.72	5.10	3.13	2.81	3.36	2.30	4.37
23 GOVERNMENT SECTOR	1.60	1.60	1.57	1.58	1.56	1.53	1.56
TOTAL: ALL SECTORS	3.01	4.61	2.58	2.27	2.63	1.62	3.14

(GROSS DOMESTIC PRODUCT AT FACTOR COST - REAL)

INSTITUTE FOR POLICY ANALYSIS, UNIVERSITY OF TORONTO
PRISM82A *** REVISED LO-TREND BASE CASE *** SEPT/82

Appendix Table 12 (cont'd.)

CANADA: REAL DOMESTIC PRODUCT BY INDUSTRY

(YEAR-OVER-YEAR PER-CENT CHANGE)

	1981	1982	1983	1984	1985	1986	1987	1988
1 AGRICULTURE, FISHING & TRAPPING	9.94	-4.24	2.32	0.51	1.63	0.79	0.80	1.75
2 FORESTRY	-2.99	-8.07	1.57	0.67	4.84	1.83	1.08	2.30
3 MINERAL FUEL MINES & WELLS	-8.01	3.17	3.22	0.20	0.53	2.76	2.75	1.18
4 OTHER MINES & QUARRIES	-4.01	-9.35	0.21	-1.12	3.33	0.13	-0.53	1.21
5 FOOD, FEED, BEVERAGES & TOBACCO	4.27	-4.19	2.63	1.94	1.61	0.76	0.63	1.82
6 TEXTILES & CLOTHING	1.23	-7.29	2.19	4.92	4.93	3.41	0.54	2.47
7 WOOD & FURNITURE	-1.05	-9.59	1.26	0.87	5.48	1.58	0.68	2.01
8 PAPER, PRINTING & ALLIED INDUSTRIES	0.64	-5.19	2.43	0.75	3.80	2.42	1.92	2.71
9 PRIMARY METAL & METAL FABRICATING	2.90	-9.28	1.01	1.91	4.80	1.55	-0.20	2.23
10 MOTOR VEHICLES & PARTS	4.97	-7.62	-1.48	20.08	7.12	2.56	-2.13	0.41
11 MACHINERY & OTH. TRANSPORTATION EQUIP.	3.87	-13.18	0.72	-0.39	2.77	4.53	1.96	3.65
12 ELECTRICAL PRODUCTS	8.60	-16.02	0.46	1.63	3.79	2.61	0.21	1.82
13 CHEMICAL, RUBBER & PETROLEUM PRODUCTS	2.83	-4.76	3.92	4.83	5.28	3.39	1.96	4.10
14 NON-METALLIC MINERAL PRODUCTS	3.52	-10.84	1.13	4.83	8.10	0.62	-1.77	2.72
15 OTHER MANUFACTURING INDUSTRIES	7.26	-10.58	2.16	2.01	3.03	2.25	0.03	1.94
TOTAL MANUFACTURING	3.20	-8.23	1.67	3.51	4.34	2.33	0.57	2.43
16 CONSTRUCTION	7.90	-12.88	0.32	5.47	9.71	-0.59	-2.94	2.03
17 ELECTRIC POWER, WATER & GAS UTILITIES	4.63	-3.16	5.21	5.64	5.58	4.43	3.46	4.45
TOTAL GOODS-PRODUCING INDUSTRIES	3.69	-7.92	1.83	3.43	4.92	1.88	0.37	2.44
18 TRANSPORTATION & STORAGE	2.35	-7.02	2.69	4.80	4.44	3.62	1.46	1.97
19 COMMUNICATION	8.07	-1.67	6.70	8.01	8.59	7.70	6.58	7.03
20 TRADE	2.36	-8.07	3.19	4.40	5.59	4.07	3.15	3.62
21 FINANCE, INSURANCE & REAL ESTATE	4.39	-2.40	3.64	4.95	5.68	4.38	3.68	3.76
22 OTHER SERVICE INDUSTRIES	7.75	-2.39	3.45	5.46	5.53	4.82	3.83	4.11
TOTAL SERVICE INDUSTRIES	4.85	-4.31	3.64	5.23	5.75	4.67	3.63	3.96
23 GOVERNMENT SECTOR	1.52	1.75	1.49	1.55	1.37	1.41	1.46	1.57
TOTAL: ALL SECTORS	4.03	-4.96	2.72	4.14	4.93	3.30	2.24	3.17

(GROSS DOMESTIC PRODUCT AT FACTOR COST - REAL)

INSTITUTE FOR POLICY ANALYSIS, UNIVERSITY OF TORONTO
PRISM82A *** REVISED LO-TREND BASE CASE *** SEPT/82

Appendix Table 12 (cont'd.)
CANADA: REAL DOMESTIC PRODUCT BY INDUSTRY

(MILLIONS OF 1971 DOLLARS)

	1989	1990	1991	1992	1993	1994	1995
1 AGRICULTURE,FISHING & TRAPPING	3489	3570	3573	3592	3643	3638	3645
2 FORESTRY	782	819	830	845	863	871	879
3 MINERAL FUEL MINES & WELLS	1623	1651	1676	1775	1798	1887	2007
4 OTHER MINES & QUARRIES	1742	1807	1813	1830	1846	1835	1821
5 FOOD,FEED,BEVERAGES & TOBACCO	3787	3869	3876	3881	3934	3925	3959
6 TEXTILES & CLOTHING	2031	2077	2082	2066	2056	2024	2016
7 WOOD & FURNITURE	1671	1756	1782	1815	1854	1874	1898
8 PAPER,PRINTING & ALLIED INDUSTRIES	4031	4188	4242	4328	4427	4472	4512
9 PRIMARY, METAL & METAL FABRICATING	4496	4717	4795	4856	4908	4877	4872
10 MOTOR VEHICLES & PARTS	2428	2524	2652	2655	2689	2690	2643
11 MACHINERY & OTH. TRANSPORTATION EQUIP.	2591	2711	2771	2807	2831	2786	2775
12 ELECTRICAL PRODUCTS	1665	1717	1728	1719	1715	1686	1690
13 CHEMICAL,RUBBER & PETROLEUM PRODUCTS	3670	3856	3952	4033	4133	4154	4233
14 NON-METALLIC MINERAL PRODUCTS	1057	1138	1169	1191	1207	1204	1227
15 OTHER MANUFACTURING INDUSTRIES	1008	1016	1000	976	953	911	897
TOTAL MANUFACTURING	28435	29567	30051	30328	30707	30605	30720
16 CONSTRUCTION	7432	8120	8407	8612	8785	8881	9187
17 ELECTRIC POWER,WATER & GAS UTILITIES	5241	5536	5737	5931	6195	6387	6660
TOTAL GOODS-PRODUCING INDUSTRIES	48746	51071	52086	52912	53837	54104	54919
18 TRANSPORTATION & STORAGE	8895	9260	9416	9569	9788	9875	10023
19 COMMUNICATION	8688	9391	9971	10552	11205	11782	12611
20 TRADE	17998	18813	19293	19690	20294	20751	21425
21 FINANCE,INSURANCE & REAL ESTATE	20698	21639	22203	22712	23284	23629	24540
22 OTHER SERVICE INDUSTRIES	22823	24036	24859	25630	26543	27172	28678
TOTAL SERVICE INDUSTRIES	79100	83138	85741	88152	91115	93208	97277
23 GOVERNMENT SECTOR	15477	15725	15972	16225	16479	16731	16992
TOTAL: ALL SECTORS	143323	149935	153799	157290	161431	164043	169189

(GROSS DOMESTIC PRODUCT AT FACTOR COST - REAL)

INSTITUTE FOR POLICY ANALYSIS, UNIVERSITY OF TORONTO
PRISM82A *** REVISED LO-TREND BASE CASE *** SEPT/82

Appendix Table 13

CANADA: EMPLOYMENT BY INDUSTRY

(THOUSANDS)

	1981	1982	1983	1984	1985	1986	1987	1988
1 AGRICULTURE, FISHING & TRAPPING	524	509	503	494	495	491	488	483
2 FORESTRY	72	67	67	67	70	72	72	73
3 MINERAL FUEL MINES & WELLS	47	50	50	50	50	51	52	52
4 OTHER MINES & QUARRIES	155	148	149	149	156	159	162	164
5 FOOD, FEED, BEVERAGES & TOBACCO	289	282	282	283	287	287	287	287
6 TEXTILES & CLOTHING	222	206	202	206	212	214	211	210
7 WOOD & FURNITURE	158	143	139	136	141	140	138	136
8 PAPER, PRINTING & ALLIED INDUSTRIES	277	266	266	263	271	275	278	280
9 PRIMARY METAL & METAL FABRICATING	321	294	288	288	299	301	297	298
10 MOTOR VEHICLES & PARTS	98	89	84	100	106	107	102	100
11 MACHINERY & OTH. TRANSPORTATION EQUIP.	234	208	200	192	191	193	191	190
12 ELECTRICAL PRODUCTS	159	135	131	129	132	133	131	130
13 CHEMICAL, RUBBER & PETROLEUM PRODUCTS	179	173	173	176	182	184	185	186
14 NON-METALLIC MINERAL PRODUCTS	62	56	55	57	62	62	60	61
15 OTHER MANUFACTURING INDUSTRIES	121	108	107	107	110	111	109	109
TOTAL MANUFACTURING	2120	1961	1928	1937	1992	2005	1991	1987
16 CONSTRUCTION	645	573	566	593	657	655	639	648
17 ELECTRIC POWER, WATER & GAS UTILITIES	119	117	118	121	125	127	130	131
TOTAL GOODS-PRODUCING INDUSTRIES	3677	3420	3376	3406	3541	3555	3529	3535
18 TRANSPORTATION & STORAGE	519	483	479	489	503	512	511	507
19 COMMUNICATION	266	257	259	268	282	293	302	309
20 TRADE	1875	1764	1785	1842	1947	2022	2089	2139
21 FINANCE, INSURANCE & REAL ESTATE	592	588	596	617	650	675	698	713
22 OTHER SERVICE INDUSTRIES	2749	2799	2838	2944	3091	3220	3345	3437
TOTAL SERVICE INDUSTRIES	6005	5897	5956	6159	6473	6721	6945	7104
23 GOVERNMENT SECTOR	1250	1268	1279	1302	1326	1350	1374	1399
TOTAL: ALL SECTORS	10933	10584	10616	10872	11345	11631	11853	12043

INSTITUTE FOR POLICY ANALYSIS, UNIVERSITY OF TORONTO
PRISM82A *** REVISED LO-TREND BASE CASE *** SEPT/82
Appendix Table 13 (cont'd.)

CANADA: EMPLOYMENT BY INDUSTRY
(YEAR-OVER-YEAR PER-CENT CHANGE)

	1989	1990	1991	1992	1993	1994	1995
1 AGRICULTURE,FISHING & TRAPPING	-0.97	0.02	-2.29	-1.51	-0.17	-1.27	-2.54
2 FORESTRY	1.62	4.15	-2.23	1.38	2.23	1.18	-0.39
3 MINERAL FUEL MINES & WELLS	-0.91	0.83	0.31	4.12	1.06	4.35	4.01
4 OTHER MINES & QUARRIES	2.10	4.38	0.88	1.91	2.32	1.19	-0.72
5 FOOD,BEVERAGES & TOBACCO	-0.16	0.90	-1.42	-1.14	0.78	-0.63	-1.10
6 TEXTILES & CLOTHING	-1.54	-0.60	-2.82	-2.68	-3.38	-3.24	-3.90
7 WOOD & FURNITURE	-0.74	2.18	-1.69	-0.75	0.12	-0.70	-2.23
8 PAPER,PRINTING & ALLIED INDUSTRIES	0.95	2.40	-0.61	0.55	1.36	0.36	-1.44
9 PRIMARY METAL & METAL FABRICATING	-0.08	3.23	-0.43	-0.24	0.02	-1.47	-2.67
10 MOTOR VEHICLES & PARTS	-3.44	1.44	2.57	-2.50	-0.73	-1.77	-5.75
11 MACHINERY & OTH. TRANSPORTATION EQUIP.	-4.22	0.21	-1.85	-2.64	-1.94	-2.87	-4.22
12 ELECTRICAL PRODUCTS	-2.58	0.45	-2.04	-2.64	-1.94	-2.87	-2.86
13 CHEMICAL,RUBBER & PETROLEUM PRODUCTS	0.50	2.23	-0.32	-0.27	0.62	-0.81	-1.25
14 NON-METALLIC MINERAL PRODUCTS	6.60	6.90	1.29	1.03	-0.41	-0.10	0.04
15 OTHER MANUFACTURING INDUSTRIES	-2.47	-1.19	-4.05	-4.41	-4.22	-5.84	-4.64
TOTAL MANUFACTURING	-0.73	1.52	-1.14	-1.21	-0.39	-1.57	-2.49
16 CONSTRUCTION	3.11	9.31	3.13	2.79	3.12	2.14	2.87
17 ELECTRIC POWER,WATER & GAS UTILITIES	1.23	2.66	0.59	0.82	2.38	1.46	0.86
TOTAL GOODS-PRODUCING INDUSTRIES	0.19	3.01	-0.23	-0.06	0.70	-0.32	-0.93
18 TRANSPORTATION & STORAGE	-0.58	1.67	-1.09	-0.66	0.61	-0.55	-1.59
19 COMMUNICATION	1.94	3.68	1.41	1.51	2.50	1.70	2.00
20 TRADE	1.92	3.69	1.46	1.47	3.09	2.60	1.83
21 FINANCE,INSURANCE & REAL ESTATE	1.73	3.06	1.99	1.13	0.79	1.32	1.71
22 OTHER SERVICE INDUSTRIES	2.68	3.95	2.31	2.64	3.65	3.13	3.68
TOTAL SERVICE INDUSTRIES	2.09	3.61	1.63	1.87	3.07	2.50	2.54
23 GOVERNMENT SECTOR	1.81	1.81	1.81	1.81	1.81	1.81	1.81
TOTAL: ALL SECTORS	1.50	3.23	1.11	1.31	2.26	1.64	1.51

INSTITUTE FOR POLICY ANALYSIS, UNIVERSITY OF TORONTO
PRISM82A *** REVISED LO-TREND BASE CASE *** SEPT/82

Appendix Table 13 (cont'd.)
CANADA: EMPLOYMENT BY INDUSTRY

(YEAR-OVER-YEAR PER-CENT CHANGE)

	1981	1982	1983	1984	1985	1986	1987	1988
1 AGRICULTURE,FISHING & TRAPPING	2.43	-2.85	-1.18	-1.83	0.27	-0.89	-0.57	-0.91
2 FORESTRY	-3.51	-6.37	-0.32	-0.46	5.22	1.70	1.25	1.24
3 MINERAL FUEL MINES & WELLS	14.02	5.17	0.74	-0.54	0.82	2.06	2.34	-0.04
4 OTHER MINES & QUARRIES	2.73	-4.22	0.31	-0.09	5.11	1.80	1.48	1.77
5 FOOD,FEED,BEVERAGES & TOBACCO	0.16	-2.63	0.17	0.44	1.21	-0.06	0.13	0.13
6 TEXTILES & CLOTHING	0.76	-6.97	-1.83	1.63	2.93	1.12	-1.30	-0.73
7 WOOD & FURNITURE	-4.36	-9.48	-2.76	-2.38	3.43	-0.73	-1.28	-1.24
8 PAPER,PRINTING & ALLIED INDUSTRIES	1.89	-3.87	-0.28	-1.13	3.17	1.39	1.20	0.76
9 PRIMARY METAL & METAL FABRICATING	-1.91	-8.39	-1.87	-0.19	4.00	0.37	-1.07	0.13
10 MOTOR VEHICLES & PARTS	0.94	-8.95	-5.85	19.79	6.01	0.66	-4.30	-2.82
11 MACHINERY & OTH. TRANSPORTATION EQUIP.	2.14	-11.29	-3.86	-3.90	-0.33	0.75	-0.82	-0.83
12 ELECTRICAL PRODUCTS	4.87	-15.05	-3.12	-1.25	2.01	0.67	-1.21	-1.04
13 CHEMICAL,RUBBER & PETROLEUM PRODUCTS	4.27	-3.32	0.05	1.73	3.31	1.29	0.39	1.00
14 NON-METALLIC MINERAL PRODUCTS	-1.98	-9.94	-1.39	3.24	8.01	-0.14	-2.24	1.09
15 OTHER MANUFACTURING INDUSTRIES	1.29	-10.41	-0.83	-0.17	2.14	1.00	-1.01	-0.26
TOTAL MANUFACTURING	0.69	-7.52	-1.66	0.49	2.82	0.62	-0.68	-0.20
16 CONSTRUCTION	4.14	-11.15	-1.26	4.89	10.78	-0.36	-2.39	1.39
17 ELECTRIC POWER,WATER & GAS UTILITIES	3.25	-2.02	1.05	2.38	3.52	2.13	1.63	1.22
TOTAL GOODS-PRODUCING INDUSTRIES	1.76	-7.00	-1.29	0.89	3.98	0.38	-0.72	0.16
18 TRANSPORTATION & STORAGE	-1.11	-6.89	-0.96	2.10	2.95	1.77	-0.10	-0.88
19 COMMUNICATION	1.92	-3.51	1.08	3.37	5.21	3.89	3.08	2.22
20 TRADE	2.39	-5.87	1.15	3.22	5.68	3.86	3.32	2.41
21 FINANCE,INSURANCE & REAL ESTATE	-2.57	-0.68	1.27	3.51	5.48	3.75	3.45	2.08
22 OTHER SERVICE INDUSTRIES	6.69	1.85	1.38	3.72	5.02	4.16	3.88	2.76
TOTAL SERVICE INDUSTRIES	3.44	-1.80	1.01	3.40	5.11	3.83	3.33	2.30
23 GOVERNMENT SECTOR	1.00	1.41	0.87	1.81	1.81	1.81	1.81	1.81
TOTAL: ALL SECTORS	2.59	-3.19	0.30	2.41	4.36	2.52	1.91	1.60

INSTITUTE FOR POLICY ANALYSIS, UNIVERSITY OF TORONTO
PRISM82A *** REVISED LO-TREND BASE CASE *** SEPT/82

Appendix Table 13 (cont'd.)

CANADA: EMPLOYMENT BY INDUSTRY

(THOUSANDS)

	1989	1990	1991	1992	1993	1994	1995
1 AGRICULTURE,FISHING & TRAPPING	479	479	468	461	460	454	443
2 FORESTRY	75	78	78	79	81	82	81
3 MINERAL FUEL MINES & WELLS	52	52	53	55	55	58	60
4 OTHER MINES & QUARRIES	168	175	177	180	184	186	185
5 FOOD,FEED,BEVERAGES & TOBACCO	287	290	285	282	284	283	279
6 TEXTILES & CLOTHING	207	205	200	193	188	182	174
7 WOOD & FURNITURE	135	138	136	135	135	134	131
8 PAPER,PRINTING & ALLIED INDUSTRIES	283	290	288	289	293	294	290
9 PRIMARY METAL & METAL FABRICATING	297	307	306	305	305	301	293
10 MOTOR VEHICLES & PARTS	96	97	100	97	97	95	90
11 MACHINERY & OTH. TRANSPORTATION EQUIP.	185	186	182	178	175	169	162
12 ELECTRICAL PRODUCTS	126	127	124	121	119	115	112
13 CHEMICAL,RUBBER & PETROLEUM PRODUCTS	187	192	191	190	192	190	188
14 NON-METALLIC MINERAL PRODUCTS	62	66	67	68	68	68	68
15 OTHER MANUFACTURING INDUSTRIES	107	105	101	97	92	87	83
TOTAL MANUFACTURING	1972	2002	1980	1956	1948	1918	1870
16 CONSTRUCTION	668	731	753	774	799	816	839
17 ELECTRIC POWER,WATER & GAS UTILITIES	133	136	137	138	141	143	145
TOTAL GOODS-PRODUCING INDUSTRIES	3542	3649	3640	3638	3664	3652	3618
18 TRANSPORTATION & STORAGE	504	512	507	504	507	504	496
19 COMMUNICATION	315	326	331	336	344	350	357
20 TRADE	2180	2261	2294	2327	2399	2462	2507
21 FINANCE,INSURANCE & REAL ESTATE	725	747	753	762	777	787	800
22 OTHER SERVICE INDUSTRIES	3529	3668	3753	3852	3993	4117	4269
TOTAL SERVICE INDUSTRIES	7253	7515	7638	7780	8020	8220	8429
23 GOVERNMENT SECTOR	1424	1450	1477	1503	1530	1558	1586
TOTAL: ALL SECTORS	12224	12619	12759	12927	13219	13435	13639

INSTITUTE FOR POLICY ANALYSIS, UNIVERSITY OF TORONTO
PRISM82A *** REVISED LO-TREND BASE CASE *** SEPT/82

Appendix Table 14
CANADA: OUTPUT PER EMPLOYEE BY INDUSTRY

(THOUSANDS OF 1971 DOLLARS)

	1981	1982	1983	1984	1985	1986	1987	1988
1 AGRICULTURE, FISHING & TRAPPING	6	6	6	7	7	7	7	7
2 FORESTRY	10	10	10	10	10	10	10	10
3 MINERAL FUEL MINES & WELLS	30	29	30	30	30	30	31	31
4 OTHER MINES & QUARRIES	12	11	11	11	11	11	10	10
5 FOOD, FEED, BEVERAGES & TOBACCO	12	12	12	13	13	13	13	13
6 TEXTILES & CLOTHING	8	8	8	9	9	9	9	10
7 WOOD & FURNITURE	10	10	11	11	11	11	12	12
8 PAPER, PRINTING & ALLIED INDUSTRIES	13	13	13	13	13	14	14	14
9 PRIMARY METAL & METAL FABRICATING	14	13	14	14	14	14	15	15
10 MOTOR VEHICLES & PARTS	21	21	22	22	23	23	24	24
11 MACHINERY & OTH. TRANSPORTATION EQUIP.	11	11	11	12	12	13	13	13
12 ELECTRICAL PRODUCTS	11	11	11	12	12	12	12	13
13 CHEMICAL, RUBBER & PETROLEUM PRODUCTS	17	16	17	17	18	18	18	19
14 NON-METALLIC MINERAL PRODUCTS	16	16	16	16	16	16	17	17
15 OTHER MANUFACTURING INDUSTRIES	8	8	9	9	9	9	9	9
TOTAL MANUFACTURING	12	12	13	13	13	14	14	14
16 CONSTRUCTION	11	11	11	11	11	11	11	11
17 ELECTRIC POWER, WATER & GAS UTILITIES	33	33	34	35	36	36	37	38
TOTAL GOODS-PRODUCING INDUSTRIES	12	12	12	13	13	13	13	13
18 TRANSPORTATION & STORAGE	15	15	15	16	16	16	17	17
19 COMMUNICATION	20	21	22	23	23	24	25	26
20 TRADE	8	8	8	8	8	8	8	8
21 FINANCE, INSURANCE & REAL ESTATE	27	26	27	27	27	28	28	28
22 OTHER SERVICE INDUSTRIES	6	6	6	6	6	6	6	6
TOTAL SERVICE INDUSTRIES	10	10	10	10	10	11	11	11
23 GOVERNMENT SECTOR	11	11	11	11	11	11	11	11
TOTAL: ALL SECTORS (GROSS DOMESTIC PRODUCT AT FACTOR COST - REAL)	11	11	11	11	11	11	11	12

INSTITUTE FOR POLICY ANALYSIS, UNIVERSITY OF TORONTO
PRISM82A * REVISED LO-TREND BASE CASE *** SEPT/82**

Appendix Table 14 (cont'd.)
CANADA: OUTPUT PER EMPLOYEE BY INDUSTRY
(YEAR-OVER-YEAR PER-CENT CHANGE)

	1989	1990	1991	1992	1993	1994	1995
1 AGRICULTURE,FISHING & TRAPPING	2.60	2.29	2.42	2.08	1.59	1.14	2.79
2 FORESTRY	0.97	0.49	1.03	0.46	0.09	-0.29	1.42
3 MINERAL FUEL MINES & WELLS	0.92	0.89	1.20	1.72	0.23	0.58	2.26
4 OTHER MINES & QUARRIES	-0.51	-0.63	-0.56	-0.93	-1.41	-1.74	-0.09
5 FOOD,FEED,BEVERAGES & TOBACCO	1.63	1.24	1.55	1.26	0.60	0.40	1.99
6 TEXTILES & CLOTHING	2.02	2.89	3.16	2.69	2.23	1.78	3.61
7 WOOD & FURNITURE	3.23	2.82	3.22	2.63	2.05	1.81	3.57
8 PAPER,PRINTING & ALLIED INDUSTRIES	1.87	1.46	1.93	1.45	0.92	0.67	2.37
9 PRIMARY METAL & METAL FABRICATING	2.09	1.63	2.09	1.52	1.05	0.85	2.63
10 MOTOR VEHICLES & PARTS	3.46	2.96	3.46	2.70	2.02	1.85	4.21
11 MACHINERY & OTH. TRANSPORTATION EQUIP.	3.05	4.40	4.16	3.40	2.91	2.85	3.96
12 ELECTRICAL PRODUCTS	2.83	2.63	2.78	2.17	1.73	1.23	3.17
13 CHEMICAL,RUBBER & PETROLEUM PRODUCTS	2.96	2.78	2.83	2.32	1.85	1.31	3.19
14 NON-METALLIC MINERAL PRODUCTS	1.55	0.98	1.49	0.82	0.23	0.20	1.83
15 OTHER MANUFACTURING INDUSTRIES	2.39	2.02	2.56	2.16	1.88	1.58	3.22
TOTAL MANUFACTURING	2.61	2.42	2.81	2.16	1.65	1.25	2.94
16 CONSTRUCTION	0.51	-0.05	0.39	-0.34	-1.08	-1.02	0.56
17 ELECTRIC POWER,WATER & GAS UTILITIES	3.13	2.89	3.01	2.55	2.02	1.61	3.39
TOTAL GOODS-PRODUCING INDUSTRIES	2.12	1.70	2.22	1.65	1.04	0.82	2.46
18 TRANSPORTATION & STORAGE	2.82	2.39	2.80	2.31	1.67	1.44	3.14
19 COMMUNICATION	4.69	4.25	4.70	4.26	3.61	3.38	4.93
20 TRADE	1.13	0.81	1.08	0.58	-0.02	-0.34	1.39
21 FINANCE,INSURANCE & REAL ESTATE	1.67	1.44	1.80	1.15	0.52	0.16	2.11
22 OTHER SERVICE INDUSTRIES	1.25	1.31	1.10	0.45	-0.08	-0.74	1.80
TOTAL SERVICE INDUSTRIES	1.59	1.44	1.48	0.92	0.28	-0.20	1.78
23 GOVERNMENT SECTOR	-0.21	-0.20	-0.24	-0.22	-0.25	-0.28	-0.24
TOTAL: ALL SECTORS	1.48	1.34	1.45	0.95	0.37	-0.02	1.60

(GROSS DOMESTIC PRODUCT AT FACTOR COST - REAL)

INSTITUTE FOR POLICY ANALYSIS, UNIVERSITY OF TORONTO
PRISM82A *** REVISED LO-TREND BASE CASE *** SEPT/82

Appendix Table 14 (cont'd.)

CANADA: OUTPUT PER EMPLOYEE BY INDUSTRY

(THOUSANDS OF 1971 DOLLARS)

	1989	1990	1991	1992	1993	1994	1995
1 AGRICULTURE, FISHING & TRAPPING	7	7	8	8	8	8	8
2 FORESTRY	10	11	11	11	11	11	11
3 MINERAL FUEL MINES & WELLS	31	31	32	32	32	33	33
4 OTHER MINES & QUARRIES	10	10	10	10	10	10	10
5 FOOD, FEED, BEVERAGES & TOBACCO	13	13	14	14	14	14	14
6 TEXTILES & CLOTHING	10	10	10	11	11	11	12
7 WOOD & FURNITURE	12	13	13	13	14	14	14
8 PAPER, PRINTING & ALLIED INDUSTRIES	14	14	15	15	15	15	16
9 PRIMARY METAL & METAL FABRICATING	15	15	16	16	16	16	17
10 MOTOR VEHICLES & PARTS	25	26	27	27	28	28	30
11 MACHINERY & OTH. TRANSPORTATION EQUIP.	14	15	15	16	16	16	17
12 ELECTRICAL PRODUCTS	13	14	14	14	14	15	15
13 CHEMICAL, RUBBER & PETROLEUM PRODUCTS	20	20	21	21	22	22	23
14 NON-METALLIC MINERAL PRODUCTS	17	17	17	18	18	18	18
15 OTHER MANUFACTURING INDUSTRIES	9	10	10	10	10	10	11
TOTAL MANUFACTURING	14	15	15	16	16	16	16
16 CONSTRUCTION	11	11	11	11	11	11	11
17 ELECTRIC POWER, WATER & GAS UTILITIES	40	41	42	43	44	45	46
TOTAL GOODS-PRODUCING INDUSTRIES	14	14	14	15	15	15	15
18 TRANSPORTATION & STORAGE	18	18	19	19	19	20	20
19 COMMUNICATION	28	29	30	31	33	34	35
20 TRADE	8	8	8	8	8	8	9
21 FINANCE, INSURANCE & REAL ESTATE	29	29	29	30	30	30	31
22 OTHER SERVICE INDUSTRIES	6	7	7	7	7	7	7
TOTAL SERVICE INDUSTRIES	11	11	11	11	11	11	12
23 GOVERNMENT SECTOR	11	11	11	11	11	11	11
TOTAL: ALL SECTORS (GROSS DOMESTIC PRODUCT AT FACTOR COST - REAL)	12	12	12	12	12	12	12

INSTITUTE FOR POLICY ANALYSIS, UNIVERSITY OF TORONTO
PRISM82A * REVISED LO-TREND BASE CASE *** SEPT/82**

Appendix Table 14 (cont'd.)

CANADA: OUTPUT PER EMPLOYEE BY INDUSTRY
(YEAR-OVER-YEAR PER-CENT CHANGE)

	1981	1982	1983	1984	1985	1986	1987	1988
1 AGRICULTURE,FISHING & TRAPPING	7.33	1.43	3.55	2.38	1.36	1.70	1.38	2.69
2 FORESTRY	0.53	-1.81	1.89	1.14	-0.35	0.13	-0.17	1.04
3 MINERAL FUEL MINES & WELLS	-19.32	-1.90	2.45	0.74	-0.29	0.69	0.41	1.23
4 OTHER MINES & QUARRIES	-6.95	-5.35	-0.10	-1.02	-1.69	-1.64	-1.98	-0.54
5 FOOD,FEED,BEVERAGES & TOBACCO	4.11	-1.60	2.45	1.49	0.59	0.83	0.50	1.68
6 TEXTILES & CLOTHING	0.47	-0.35	4.09	3.23	1.94	2.26	1.87	3.22
7 WOOD & FURNITURE	3.45	-0.12	4.14	3.33	1.98	2.32	1.98	3.29
8 PAPER,PRINTING & ALLIED INDUSTRIES	-1.23	-1.38	2.72	1.90	0.60	1.02	0.71	1.93
9 PRIMARY METAL & METAL FABRICATING	4.90	-0.98	2.94	2.10	0.77	1.18	0.88	2.10
10 MOTOR VEHICLES & PARTS	4.00	1.47	4.64	0.25	1.05	1.68	2.27	3.32
11 MACHINERY & OTH. TRANSPORTATION EQUIP.	1.70	-2.13	4.76	3.65	3.11	3.74	2.81	4.51
12 ELECTRICAL PRODUCTS	3.56	-1.14	3.69	2.92	1.74	1.93	1.44	2.90
13 CHEMICAL,RUBBER & PETROLEUM PRODUCTS	-1.38	1.49	3.87	3.04	2.07	2.07	1.56	3.07
14 NON-METALLIC MINERAL PRODUCTS	5.61	-1.00	2.56	1.55	0.09	0.76	0.47	1.61
15 OTHER MANUFACTURING INDUSTRIES	5.89	-0.18	3.01	2.19	0.87	1.24	1.05	2.20
TOTAL MANUFACTURING	2.48	-0.76	3.38	3.01	1.48	1.69	1.26	2.63
16 CONSTRUCTION	3.61	-1.94	1.60	0.55	-0.97	-0.23	-0.57	0.64
17 ELECTRIC POWER,WATER & GAS UTILITIES	1.33	-1.16	4.12	3.18	1.99	2.26	1.81	3.20
TOTAL GOODS-PRODUCING INDUSTRIES	1.90	-0.99	3.16	2.52	0.90	1.50	1.10	2.27
18 TRANSPORTATION & STORAGE	3.50	-0.15	3.69	2.65	1.45	1.82	1.56	2.87
19 COMMUNICATION	6.04	1.91	5.56	4.49	3.21	3.66	3.39	4.70
20 TRADE	-0.03	-2.33	2.02	1.15	-0.08	0.20	0.16	1.18
21 FINANCE, INSURANCE & REAL ESTATE	7.14	-1.73	2.34	1.40	0.20	0.61	0.22	1.64
22 OTHER SERVICE INDUSTRIES	0.99	-4.16	2.04	1.68	0.49	0.64	-0.05	1.32
TOTAL SERVICE INDUSTRIES	1.36	-2.55	2.60	1.77	0.61	0.81	0.29	1.63
23 GOVERNMENT SECTOR	0.51	0.34	0.62	-0.26	-0.43	-0.40	-0.34	-0.24
TOTAL: ALL SECTORS	1.40	-1.83	2.42	1.69	0.55	0.77	0.32	1.54

(GROSS DOMESTIC PRODUCT AT FACTOR COST - REAL)

INSTITUTE FOR POLICY ANALYSIS, UNIVERSITY OF TORONTO
PRISM82A *** REVISED LO-TREND BASE CASE *** SEPT/82

Appendix Table 15.1

SUMMARY - PROVINCIAL ECONOMY

NEWFOUNDLAND

	1981	1982	1983	1984	1985	1986	1987	1988
PROVINCIAL GROSS DOMESTIC PRODUCT	4360	4800	5479	6398	7505	8582	10890	13523
REAL PROVINCIAL GDP AT FACTOR COST	1669	1611	1667	1752	1845	1905	2002	2101
DEFLATOR, PROV'L GDP AT FACTOR COST (71=1.0)	2.45	2.76	3.04	3.37	3.78	4.17	4.98	5.88
REAL PERSONAL DISP. INCOME PER CAP. ($'000)	2.506	2.585	2.636	2.732	2.793	2.836	2.920	3.004
ANNUAL WAGES & SALARIES PER EMPLOYEE ($'000)	13.626	15.318	16.279	17.992	19.552	21.540	24.300	26.984
GDP AT FACTOR COST PER EMPLOYEE ($71 '000)	8.915	8.788	8.963	9.076	9.099	9.121	9.262	9.500
LABOUR FORCE ('000)	218	219	225	233	241	247	255	258
EMPLOYMENT ('000)	187	183	186	193	203	209	216	221
UNEMPLOYMENT RATE (%)	14.16	16.38	17.36	17.18	15.91	15.60	15.14	14.31
REAL DOMESTIC PRODUCT - MANUFACTURING	159	151	155	158	164	167	170	175
RDP - GOODS-PRODUCING INDUSTRIES	670	635	662	702	742	757	813	865
RDP - SERVICE INDUSTRIES	726	696	721	761	808	848	883	924

NOTE - NOMINAL VALUES IN MILLIONS OF CURRENT DOLLARS
     - REAL VALUES IN MILLIONS OF 1971 DOLLARS

INSTITUTE FOR POLICY ANALYSIS, UNIVERSITY OF TORONTO
PRISM82A * REVISED LO-TREND BASE CASE *** SEPT/82**

Appendix Table 15.1 (cont'd.)

SUMMARY - PROVINCIAL ECONOMY

NEWFOUNDLAND

	1989	1990	1991	1992	1993	1994	1995
PROVINCIAL GROSS DOMESTIC PRODUCT	15641	17897	20125	22948	24936	28052	31605
REAL PROVINCIAL GDP AT FACTOR COST	2203	2301	2369	2438	2485	2538	2616
DEFLATOR, PROV'L GDP AT FACTOR COST (71=1.0)	6.62	7.26	7.94	8.70	9.24	10.12	11.07
REAL PERSONAL DISP. INCOME PER CAP. ($'000)	3.097	3.173	3.236	3.272	3.329	3.370	3.460
ANNUAL WAGES & SALARIES PER EMPLOYEE ($'000)	29.864	32.424	35.703	39.505	43.153	47.810	52.837
GDP AT FACTOR COST PER EMPLOYEE ($71 '000)	9.670	9.750	9.848	9.920	9.859	9.842	9.951
LABOUR FORCE ('000)	264	272	275	281	287	293	298
EMPLOYMENT ('000)	228	236	241	246	252	258	263
UNEMPLOYMENT RATE (%)	13.87	13.22	12.65	12.53	12.27	12.07	11.80
REAL DOMESTIC PRODUCT - MANUFACTURING	179	186	188	191	195	196	198
RDP - GOODS-PRODUCING INDUSTRIES	916	949	973	1008	1028	1052
RDP - SERVICE INDUSTRIES	966	1024	1061	1094	1131	1158	1206

NOTE - NOMINAL VALUES IN MILLIONS OF CURRENT DOLLARS
       - REAL VALUES IN MILLIONS OF 1971 DOLLARS

INSTITUTE FOR POLICY ANALYSIS, UNIVERSITY OF TORONTO
PRISM82A *** REVISED LO-TREND BASE CASE *** SEPT/82

Appendix Table 15.1 (cont'd.)
SUMMARY - PROVINCIAL ECONOMY

(PERCENTAGE CHANGE;   * INDICATES CHANGE IN LEVELS)

NEWFOUNDLAND

	1981	1982	1983	1984	1985	1986	1987	1988
PROVINCIAL GROSS DOMESTIC PRODUCT	17.20	10.08	14.15	16.77	17.31	14.35	26.89	24.17
REAL PROVINCIAL GDP AT FACTOR COST	3.26	-3.44	3.44	5.12	5.33	3.21	5.09	4.98
DEFLATOR, PROV'L GDP AT FACTOR COST	8.86	12.49	10.32	10.68	12.31	10.28	19.36	18.06
REAL PERSONAL DISP. INCOME PER CAP.	0.04	3.14	1.99	3.63	2.26	1.54	2.93	2.90
ANNUAL WAGES & SALARIES PER EMPLOYEE	12.85	12.42	6.27	10.52	8.67	10.17	12.81	11.04
GDP AT FACTOR COST PER EMPLOYEE	1.84	-1.42	1.99	1.26	0.26	0.23	1.55	2.57
LABOUR FORCE	2.12	0.55	2.63	3.58	3.47	2.59	2.93	1.36
EMPLOYMENT	1.39	-2.05	1.42	3.81	5.05	2.98	3.49	2.35
UNEMPLOYMENT RATE (%) *	0.62	2.22	0.99	-0.18	-1.27	-0.31	-0.46	-0.83
REAL DOMESTIC PRODUCT - MANUFACTURING	3.57	-4.94	2.68	1.95	3.41	2.02	1.61	2.82
RDP - GOODS-PRODUCING INDUSTRIES	2.92	-5.22	4.28	6.08	5.77	1.99	7.34	6.40
RDP - SERVICE INDUSTRIES	4.26	-4.04	3.51	5.52	6.24	4.90	4.16	4.63

INSTITUTE FOR POLICY ANALYSIS, UNIVERSITY OF TORONTO
PRISM82A * REVISED LO-TREND BASE CASE *** SEPT/82**

Appendix Table 15.1 (cont'd.)

SUMMARY - PROVINCIAL ECONOMY

(PERCENTAGE CHANGE; * INDICATES CHANGE IN LEVELS)

NEWFOUNDLAND

	1989	1990	1991	1992	1993	1994	1995
PROVINCIAL GROSS DOMESTIC PRODUCT	15.67	14.42	12.45	14.03	8.66	12.49	12.67
REAL PROVINCIAL GDP AT FACTOR COST	4.84	4.46	2.93	2.92	1.94	2.12	3.08
DEFLATOR, PROV'L GDP AT FACTOR COST	12.63	9.64	9.38	9.55	6.29	9.48	9.40
REAL PERSONAL DISP. INCOME PER CAP.	3.09	2.45	1.98	1.12	1.76	1.22	2.66
ANNUAL WAGES & SALARIES PER EMPLOYEE	10.67	8.57	10.11	10.65	9.23	10.79	10.51
GDP AT FACTOR COST PER EMPLOYEE	1.80	0.82	1.01	0.73	-0.61	-0.17	1.10
LABOUR FORCE	2.47	2.83	1.23	2.03	2.27	2.05	1.65
EMPLOYMENT	2.99	3.61	1.90	2.17	2.57	2.29	1.96
UNEMPLOYMENT RATE (%) *	-0.44	-0.65	-0.57	-0.12	-0.26	-0.21	-0.27
REAL DOMESTIC PRODUCT - MANUFACTURING	2.64	3.71	1.32	1.48	1.98	0.57	1.11
RDP - GOODS-PRODUCING INDUSTRIES	5.98	3.59	2.49	3.01	0.54	1.97	2.35
RDP - SERVICE INDUSTRIES	4.59	5.96	3.61	3.13	3.34	2.40	4.18

INSTITUTE FOR POLICY ANALYSIS, UNIVERSITY OF TORONTO
PRISM82A *** REVISED LO-TREND BASE CASE *** SEPT/82

Appendix Table 15.2
SUMMARY - PROVINCIAL ECONOMY

PRINCE EDWARD ISLAND

	1981	1982	1983	1984	1985	1986	1987	1988
PROVINCIAL GROSS DOMESTIC PRODUCT	918	912	1009	1144	1301	1477	1679	1892
REAL PROVINCIAL GDP AT FACTOR COST	337	302	308	317	327	336	343	353
DEFLATOR, PROV'L GDP AT FACTOR COST (71=1.0)	2.67	2.96	3.23	3.54	3.93	4.32	4.77	5.24
REAL PERSONAL DISP. INCOME PER CAP. ($'000)	3.007	2.884	2.900	2.949	2.957	2.977	3.013	3.070
ANNUAL WAGES & SALARIES PER EMPLOYEE ($'000)	11.312	12.224	12.848	14.085	15.064	16.656	18.774	20.784
GDP AT FACTOR COST PER EMPLOYEE ($71 '000)	7.058	6.630	6.764	6.860	6.882	6.928	6.951	7.047
LABOUR FORCE ('000)	54	52	53	53	54	56	57	57
EMPLOYMENT ('000)	48	45	46	46	48	48	49	50
UNEMPLOYMENT RATE (%)	11.12	13.27	13.42	13.52	12.64	12.81	12.75	12.23
REAL DOMESTIC PRODUCT - MANUFACTURING	28	26	27	28	28	29	29	30
RDP - GOODS-PRODUCING INDUSTRIES	91	62	64	66	69	70	71	73
RDP - SERVICE INDUSTRIES	149	141	145	151	158	164	170	176

NOTE - NOMINAL VALUES IN MILLIONS OF CURRENT DOLLARS
     - REAL VALUES IN MILLIONS OF 1971 DOLLARS

INSTITUTE FOR POLICY ANALYSIS, UNIVERSITY OF TORONTO
PRISM82A   *** REVISED LO-TREND BASE CASE *** SEPT/82

Appendix Table 15.2 (cont'd.)

SUMMARY - PROVINCIAL ECONOMY

PRINCE EDWARD ISLAND

	1989	1990	1991	1992	1993	1994	1995
PROVINCIAL GROSS DOMESTIC PRODUCT	2114	2389	2649	2982	3347	3746	4213
REAL PROVINCIAL GDP AT FACTOR COST	363	376	385	393	403	410	422
DEFLATOR, PROV'L GDP AT FACTOR COST (71=1.0)	5.72	6.26	6.87	7.48	8.19	8.97	9.84
REAL PERSONAL DISP. INCOME PER CAP. ($)	3,115	3,164	3,192	3,216	3,280	3,321	3,415
ANNUAL WAGES & SALARIES PER EMPLOYEE ($'000)	22,957	24,892	27,531	30,717	33,987	38,163	42,451
GDP AT FACTOR COST PER EMPLOYEE ($71,'000)	7,145	7,229	7,329	7,397	7,425	7,436	7,550
LABOUR FORCE ('000)	58	59	59	60	61	62	63
EMPLOYMENT ('000)	51	52	53	53	54	55	56
UNEMPLOYMENT RATE (%)	12.20	11.72	11.40	11.79	11.54	11.50	11.34
REAL DOMESTIC PRODUCT - MANUFACTURING	30	31	32	32	32	32	33
RDP - GOODS-PRODUCING INDUSTRIES	75	78	80	81	83	84	85
RDP - SERVICE INDUSTRIES	182	191	196	201	207	212	221

NOTE - NOMINAL VALUES IN MILLIONS OF CURRENT DOLLARS
     - REAL VALUES IN MILLIONS OF 1971 DOLLARS

INSTITUTE FOR POLICY ANALYSIS, UNIVERSITY OF TORONTO
PRISM82A *** REVISED LO-TREND BASE CASE *** SEPT/82

Appendix Table 15.2 (cont'd.)
SUMMARY - PROVINCIAL ECONOMY

(PERCENTAGE CHANGE; * INDICATES CHANGE IN LEVELS)

PRINCE EDWARD ISLAND

	1981	1982	1983	1984	1985	1986	1987	1988
PROVINCIAL GROSS DOMESTIC PRODUCT	16.71	-0.73	10.74	13.30	13.73	13.53	13.68	12.74
REAL PROVINCIAL GDP AT FACTOR COST	6.92	-10.55	2.07	3.06	3.09	2.74	2.22	2.79
DEFLATOR, PROV'L GDP AT FACTOR COST	5.58	10.72	9.01	9.81	10.91	10.04	10.36	9.78
REAL PERSONAL DISP. INCOME PER CAP.	5.50	-4.09	0.56	1.70	0.27	0.67	1.22	1.88
ANNUAL WAGES & SALARIES PER EMPLOYEE	18.06	8.06	5.10	9.63	6.95	10.57	12.71	10.71
GDP AT FACTOR COST PER EMPLOYEE	7.49	-6.06	2.01	1.42	0.32	0.66	0.34	1.38
LABOUR FORCE	-0.21	-2.41	0.24	1.73	1.72	2.25	1.81	0.80
EMPLOYMENT	-0.53	-4.77	0.06	1.62	2.76	2.06	1.88	1.39
UNEMPLOYMENT RATE (%) *	0.29	2.15	0.16	0.10	-0.88	0.16	-0.06	-0.51
REAL DOMESTIC PRODUCT - MANUFACTURING	4.14	-5.34	2.69	2.44	2.66	1.73	1.14	2.47
RDP - GOODS-PRODUCING INDUSTRIES	17.28	-31.44	2.95	2.81	3.81	1.99	1.43	2.95
RDP - SERVICE INDUSTRIES	4.82	-5.43	2.50	4.45	4.37	4.09	3.17	3.58

INSTITUTE FOR POLICY ANALYSIS, UNIVERSITY OF TORONTO
*** REVISED LO-TREND BASE CASE *** SEPT/82
PRISM82A

Appendix Table 15.2 (cont'd.)

SUMMARY - PROVINCIAL ECONOMY

PRINCE EDWARD ISLAND

(PERCENTAGE CHANGE; * INDICATES CHANGE IN LEVELS)

	1989	1990	1991	1992	1993	1994	1995
PROVINCIAL GROSS DOMESTIC PRODUCT	11.71	13.00	10.89	12.58	12.24	11.91	12.47
REAL PROVINCIAL GDP AT FACTOR COST	2.80	3.75	2.22	2.07	2.52	1.75	3.01
DEFLATOR, PROV'L GDP AT FACTOR COST	9.18	9.56	9.62	8.91	9.46	9.54	9.70
REAL PERSONAL DISP. INCOME PER CAP.	1.45	1.59	0.87	0.76	1.99	1.24	2.82
ANNUAL WAGES & SALARIES PER EMPLOYEE	10.46	8.43	10.60	11.57	10.65	12.29	11.24
GDP AT FACTOR COST PER EMPLOYEE	1.39	1.18	1.39	0.92	0.37	0.16	1.52
LABOUR FORCE	1.36	1.97	0.46	1.58	1.86	1.54	1.29
EMPLOYMENT	1.39	2.54	0.82	1.13	2.14	1.59	1.47
UNEMPLOYMENT RATE (%) *	-0.03	-0.49	-0.32	0.39	0.24	-0.04	-0.16
REAL DOMESTIC PRODUCT - MANUFACTURING	2.02	3.03	0.94	0.72	1.62	0.02	1.00
RDP - GOODS-PRODUCING INDUSTRIES	2.84	4.21	1.72	1.66	2.46	0.96	1.86
RDP - SERVICE INDUSTRIES	3.53	4.87	2.86	2.56	3.13	2.21	4.26

INSTITUTE FOR POLICY ANALYSIS, UNIVERSITY OF TORONTO
PRISM82A   *** REVISED LO-TREND BASE CASE *** SEPT/82

Appendix Table 15.3
SUMMARY - PROVINCIAL ECONOMY

NOVA SCOTIA

	1981	1982	1983	1984	1985	1986	1987	1988
PROVINCIAL GROSS DOMESTIC PRODUCT	7722	8043	9186	10694	12668	15038	17221	19620
REAL PROVINCIAL GDP AT FACTOR COST	2912	2671	2772	2908	3065	3200	3281	3398
DEFLATOR, PROV'L GDP AT FACTOR COST (71=1.0)	2.47	2.76	3.04	3.35	3.81	4.30	4.75	5.22
REAL PERSONAL DISP. INCOME PER CAP. ($'000)	3.146	3.077	3.104	3.191	3.241	3.285	3.335	3.416
ANNUAL WAGES & SALARIES PER EMPLOYEE ($'000)	14.753	16.302	17.245	18.972	20.444	22.649	25.393	28.108
GDP AT FACTOR COST PER EMPLOYEE ($71 '000)	8.832	8.565	8.788	8.950	9.050	9.176	9.217	9.362
LABOUR FORCE ('000)	367	361	365	373	383	393	402	408
EMPLOYMENT ('000)	330	312	315	325	339	349	356	363
UNEMPLOYMENT RATE (%)	10.17	13.56	13.47	12.93	11.63	11.34	11.40	11.05
REAL DOMESTIC PRODUCT - MANUFACTURING	392	337	358	381	411	425	435	453
RDP - GOODS-PRODUCING INDUSTRIES	820	719	764	820	888	931	935	965
RDP - SERVICE INDUSTRIES	1402	1252	1298	1367	1447	1525	1590	1661

NOTE - NOMINAL VALUES IN MILLIONS OF CURRENT DOLLARS
     - REAL VALUES IN MILLIONS OF 1971 DOLLARS

INSTITUTE FOR POLICY ANALYSIS, UNIVERSITY OF TORONTO
PRISM82A *** REVISED LO-TREND BASE CASE *** SEPT/82

Appendix Table 15.3 (cont'd.)

SUMMARY - PROVINCIAL ECONOMY

NOVA SCOTIA

	1989	1990	1991	1992	1993	1994	1995
PROVINCIAL GROSS DOMESTIC PRODUCT	22065	25128	28448	32682	36228	41164	46649
REAL PROVINCIAL GDP AT FACTOR COST	3516	3683	3799	3913	4010	4108	4254
DEFLATOR, PROV'L GDP AT FACTOR COST (71=1.0)	5.69	6.22	6.85	7.52	8.11	8.92	9.77
REAL PERSONAL DISP. INCOME PER CAP. ($'000)	3.473	3.547	3.593	3.635	3.707	3.762	3.887
ANNUAL WAGES & SALARIES PER EMPLOYEE ($'000)	30.998	33.685	37.217	41.380	45.475	50.724	56.278
GDP AT FACTOR COST PER EMPLOYEE ($71,'000)	9.506	9.638	9.792	9.908	9.933	9.965	10.120
LABOUR FORCE ('000)	415	426	430	438	446	455	463
EMPLOYMENT ('000)	370	382	388	395	404	412	420
UNEMPLOYMENT RATE (%)	10.91	10.29	9.80	9.73	9.54	9.40	9.20
REAL DOMESTIC PRODUCT - MANUFACTURING	469	493	507	519	534	541	553
RDP - GOODS-PRODUCING INDUSTRIES	997	1054	1089	1125	1134	1159	1188
RDP - SERVICE INDUSTRIES	1733	1829	1895	1958	2032	2088	2190

NOTE - NOMINAL VALUES IN MILLIONS OF CURRENT DOLLARS
- REAL VALUES IN MILLIONS OF 1971 DOLLARS

INSTITUTE FOR POLICY ANALYSIS, UNIVERSITY OF TORONTO
PRISM82A *** REVISED LO-TREND BASE CASE *** SEPT/82

Appendix Table 15.3 (cont'd.)
SUMMARY - PROVINCIAL ECONOMY

(PERCENTAGE CHANGE;   * INDICATES CHANGE IN LEVELS)

NOVA SCOTIA

	1981	1982	1983	1984	1985	1986	1987	1988
PROVINCIAL GROSS DOMESTIC PRODUCT	19.99	4.15	14.21	16.41	18.47	18.70	14.52	13.93
REAL PROVINCIAL GDP AT FACTOR COST	4.56	-8.26	3.77	4.92	5.38	4.42	2.54	3.54
DEFLATOR, PROV'L GDP AT FACTOR COST	9.43	12.03	9.86	10.44	13.53	12.91	10.43	9.88
REAL PERSONAL DISP. INCOME PER CAP.	1.97	-2.20	0.88	2.80	1.55	1.38	1.50	2.43
ANNUAL WAGES & SALARIES PER EMPLOYEE	14.51	10.50	5.79	10.01	7.76	10.79	12.11	10.69
GDP AT FACTOR COST PER EMPLOYEE	3.91	-3.03	2.61	1.84	1.12	1.39	0.45	1.57
LABOUR FORCE	1.08	-1.69	1.02	2.39	2.69	2.65	2.14	1.54
EMPLOYMENT	0.63	-5.40	1.13	3.03	4.22	2.98	2.08	1.94
UNEMPLOYMENT RATE (%) *	0.40	3.39	-0.10	-0.54	-1.30	-0.29	0.05	-0.35
REAL DOMESTIC PRODUCT - MANUFACTURING	5.74	-14.08	6.23	6.45	7.77	3.59	2.37	4.03
RDP - GOODS-PRODUCING INDUSTRIES	6.50	-12.38	6.28	7.38	8.22	4.93	0.34	3.24
RDP - SERVICE INDUSTRIES	4.92	-10.65	3.65	5.32	5.81	5.42	4.24	4.52

INSTITUTE FOR POLICY ANALYSIS, UNIVERSITY OF TORONTO
PRISM82A *** REVISED LO-TREND BASE CASE *** SEPT/82

Appendix Table 15.3 (cont'd.)
SUMMARY - PROVINCIAL ECONOMY
(PERCENTAGE CHANGE; * INDICATES CHANGE IN LEVELS)

NOVA SCOTIA

	1989	1990	1991	1992	1993	1994	1995
PROVINCIAL GROSS DOMESTIC PRODUCT	12.46	13.88	13.22	14.88	10.85	13.62	13.33
REAL PROVINCIAL GDP AT FACTOR COST	3.47	4.76	3.14	3.01	2.49	2.43	3.55
DEFLATOR, PROV'L GDP AT FACTOR COST	9.03	9.42	10.09	9.74	7.83	10.06	9.53
REAL PERSONAL DISP. INCOME PER CAP.	1.68	2.15	1.28	1.18	1.96	1.48	3.33
ANNUAL WAGES & SALARIES PER EMPLOYEE	10.28	8.67	10.48	11.18	9.90	11.54	10.95
GDP AT FACTOR COST PER EMPLOYEE	1.53	1.39	1.60	1.19	0.25	0.32	1.55
LABOUR FORCE	1.75	2.62	0.96	1.73	2.01	1.95	1.74
EMPLOYMENT	1.91	3.33	1.51	1.80	2.23	2.10	1.97
UNEMPLOYMENT RATE (%) *	-0.14	-0.62	-0.49	-0.06	-0.20	-0.14	-0.20
REAL DOMESTIC PRODUCT - MANUFACTURING	3.57	5.17	2.70	2.44	2.89	1.25	2.17
RDP - GOODS-PRODUCING INDUSTRIES	3.31	5.72	3.35	3.30	0.78	2.25	2.49
RDP - SERVICE INDUSTRIES	4.29	5.55	3.60	3.34	3.77	2.79	4.85

INSTITUTE FOR POLICY ANALYSIS, UNIVERSITY OF TORONTO
PRISM82A *** REVISED LO-TREND BASE CASE *** SEPT/82

Appendix Table 15.4
SUMMARY - PROVINCIAL ECONOMY

NEW BRUNSWICK

	1981	1982	1983	1984	1985	1986	1987	1988
PROVINCIAL GROSS DOMESTIC PRODUCT	6081	6615	7420	8463	9629	10915	12384	13937
REAL PROVINCIAL GDP AT FACTOR COST	2214	2118	2175	2258	2359	2431	2484	2559
DEFLATOR, PROV'L GDP AT FACTOR COST (71=1.0)	2.61	2.90	3.16	3.45	3.81	4.16	4.55	4.96
REAL PERSONAL DISP. INCOME PER CAP. ($'000)	2.841	2.852	2.863	2.921	2.960	2.983	3.020	3.080
ANNUAL WAGES & SALARIES PER EMPLOYEE ($'000)	13.946	15.430	16.311	17.868	19.197	21.084	23.497	25.859
GDP AT FACTOR COST PER EMPLOYEE ($71 '000)	8.458	8.261	8.445	8.578	8.614	8.661	8.672	8.790
LABOUR FORCE ('000)	296	299	303	311	318	325	331	336
EMPLOYMENT ('000)	262	256	258	263	274	281	286	291
UNEMPLOYMENT RATE (%)	11.64	14.33	15.13	15.25	13.96	13.77	13.60	13.24
REAL DOMESTIC PRODUCT - MANUFACTURING	338	316	324	330	343	350	354	364
RDP - GOODS-PRODUCING INDUSTRIES	687	648	660	680	710	721	724	743
RDP - SERVICE INDUSTRIES	1118	1053	1092	1149	1214	1269	1313	1363

NOTE - NOMINAL VALUES IN MILLIONS OF CURRENT DOLLARS
       - REAL VALUES IN MILLIONS OF 1971 DOLLARS

INSTITUTE FOR POLICY ANALYSIS, UNIVERSITY OF TORONTO
*** REVISED LO-TREND BASE CASE *** SEPT/82
PRISM82A

Appendix Table 15.4 (cont'd.)

SUMMARY - PROVINCIAL ECONOMY

NEW BRUNSWICK

	1989	1990	1991	1992	1993	1994	1995
PROVINCIAL GROSS DOMESTIC PRODUCT	15482	17444	19368	21941	24579	27486	30973
REAL PROVINCIAL GDP AT FACTOR COST	2637	2756	2825	2891	2970	3024	3120
DEFLATOR, PROV'L GDP AT FACTOR COST (71=1.0)	5.38	5.85	6.37	6.88	7.48	8.12	8.87
REAL PERSONAL DISP. INCOME PER CAP. ($'000)	3.121	3.180	3.206	3.225	3.288	3.317	3.409
ANNUAL WAGES & SALARIES PER EMPLOYEE ($'000)	28.368	30.724	33.751	37.258	40.838	45.231	49.923
GDP AT FACTOR COST PER EMPLOYEE ($71'000)	8.908	9.018	9.132	9.206	9.231	9.224	9.346
LABOUR FORCE ('000)	341	348	351	356	363	369	374
EMPLOYMENT ('000)	296	306	309	314	322	328	334
UNEMPLOYMENT RATE (%)	13.14	12.19	11.81	11.83	11.33	11.10	10.79
REAL DOMESTIC PRODUCT - MANUFACTURING	373	388	393	399	406	408	412
RDP - GOODS-PRODUCING INDUSTRIES	764	804	820	835	854	861	878
RDP - SERVICE INDUSTRIES	1412	1484	1530	1573	1627	1665	1736

NOTE - NOMINAL VALUES IN MILLIONS OF CURRENT DOLLARS
     - REAL VALUES IN MILLIONS OF 1971 DOLLARS

INSTITUTE FOR POLICY ANALYSIS, UNIVERSITY OF TORONTO
PRISM82A *** REVISED LO-TREND BASE CASE *** SEPT/82

Appendix Table 15.4 (cont'd.)
SUMMARY - PROVINCIAL ECONOMY

(PERCENTAGE CHANGE; * INDICATES CHANGE IN LEVELS)

NEW BRUNSWICK

	1981	1982	1983	1984	1985	1986	1987	1988
PROVINCIAL GROSS DOMESTIC PRODUCT	16.40	8.78	12.17	14.04	13.78	13.36	13.46	12.54
REAL PROVINCIAL GDP AT FACTOR COST	3.74	-4.35	2.71	3.81	4.48	3.03	2.17	3.03
DEFLATOR, PROV'L GDP AT FACTOR COST	6.56	11.37	8.88	9.24	10.42	9.21	9.38	8.97
REAL PERSONAL DISP. INCOME PER CAP.	1.51	0.37	0.41	2.00	1.35	0.77	1.25	1.97
ANNUAL WAGES & SALARIES PER EMPLOYEE	11.63	10.64	5.71	9.54	7.44	9.83	11.44	10.05
GDP AT FACTOR COST PER EMPLOYEE	1.67	-2.33	2.23	1.57	0.42	0.54	0.13	1.36
LABOUR FORCE	2.53	1.01	1.41	2.36	2.48	2.25	1.83	1.23
EMPLOYMENT	2.04	-2.06	0.47	2.21	4.04	2.48	2.04	1.65
UNEMPLOYMENT RATE (%) *	0.43	2.69	0.80	0.13	-1.29	-0.19	-0.17	-0.36
REAL DOMESTIC PRODUCT - MANUFACTURING	-0.04	-6.36	2.30	1.95	3.80	2.15	1.23	2.70
RDP - GOODS-PRODUCING INDUSTRIES	4.28	-5.61	1.82	2.97	4.42	1.51	0.49	2.61
RDP - SERVICE INDUSTRIES	4.31	-5.84	3.72	5.19	5.68	4.54	3.41	3.80

INSTITUTE FOR POLICY ANALYSIS, UNIVERSITY OF TORONTO
PRISM82A * REVISED LO-TREND BASE CASE *** SEPT/82**

Appendix Table 15.4 (cont'd.)
SUMMARY - PROVINCIAL ECONOMY

(PERCENTAGE CHANGE: * INDICATES CHANGE IN LEVELS)

NEW BRUNSWICK

	1989	1990	1991	1992	1993	1994	1995
PROVINCIAL GROSS DOMESTIC PRODUCT	11.08	12.67	11.03	13.29	12.02	11.83	12.69
REAL PROVINCIAL GDP AT FACTOR COST	3.04	4.51	2.52	2.33	2.75	1.81	3.16
DEFLATOR, PROV'L GDP AT FACTOR COST	8.34	8.85	8.85	8.08	8.58	8.65	9.20
REAL PERSONAL DISP. INCOME PER CAP.	1.32	1.91	0.83	0.58	1.94	0.90	2.76
ANNUAL WAGES & SALARIES PER EMPLOYEE	9.70	8.30	9.85	10.39	9.61	10.76	10.37
GDP AT FACTOR COST PER EMPLOYEE	1.35	1.23	1.27	0.80	0.27	-0.07	1.33
LABOUR FORCE	1.55	2.13	0.80	1.53	1.90	1.62	1.45
EMPLOYMENT	1.66	3.25	1.23	1.52	2.47	1.88	1.81
UNEMPLOYMENT RATE (%) *	-0.10	-0.95	-0.38	0.02	-0.49	-0.23	-0.32
REAL DOMESTIC PRODUCT - MANUFACTURING	2.42	4.02	1.43	1.42	1.88	0.36	1.02
RDP - GOODS-PRODUCING INDUSTRIES	2.83	5.21	2.03	1.86	2.21	0.88	1.92
RDP - SERVICE INDUSTRIES	3.64	5.09	3.09	2.81	3.40	2.37	4.26

INSTITUTE FOR POLICY ANALYSIS, UNIVERSITY OF TORONTO
PRISM82A   *** REVISED LO-TREND BASE CASE ***   SEPT/82

Appendix Table 15.5
SUMMARY - PROVINCIAL ECONOMY

QUEBEC

	1981	1982	1983	1984	1985	1986	1987	1988
PROVINCIAL GROSS DOMESTIC PRODUCT	77164	77936	87695	99796	114396	128545	143388	160380
REAL PROVINCIAL GDP AT FACTOR COST	28383	25742	26487	27467	28553	29379	29911	30719
DEFLATOR, PROV'L GDP AT FACTOR COST (71=1.0)	2.38	2.66	2.92	3.21	3.56	3.88	4.25	4.63
REAL PERSONAL DISP. INCOME PER CAP. ($'000)	3.579	3.459	3.486	3.581	3.628	3.660	3.707	3.786
ANNUAL WAGES & SALARIES PER EMPLOYEE ($'000)	16.827	18.078	19.173	20.964	22.374	24.447	27.110	29.708
GDP AT FACTOR COST PER EMPLOYEE ($71 '000)	10.571	10.041	10.301	10.476	10.539	10.618	10.652	10.821
LABOUR FORCE ('000)	2997	2988	3007	3052	3103	3157	3202	3224
EMPLOYMENT ('000)	2685	2564	2571	2622	2709	2767	2808	2839
UNEMPLOYMENT RATE (%)	10.41	14.19	14.48	14.09	12.68	12.36	12.29	11.94
REAL DOMESTIC PRODUCT - MANUFACTURING	7064	6303	6475	6628	6870	7006	7035	7187
RDP - GOODS-PRODUCING INDUSTRIES	10468	9012	9364	9723	10116	10263	10266	10472
RDP - SERVICE INDUSTRIES	14874	13665	14048	14657	15349	16000	16501	17073

NOTE - NOMINAL VALUES IN MILLIONS OF CURRENT DOLLARS
     - REAL VALUES IN MILLIONS OF 1971 DOLLARS

INSTITUTE FOR POLICY ANALYSIS, UNIVERSITY OF TORONTO
PRISM82A *** REVISED LO-TREND BASE CASE *** SEPT/82

Appendix Table 15.5 (cont'd.)
SUMMARY - PROVINCIAL ECONOMY

QUEBEC

	1989	1990	1991	1992	1993	1994	1995
PROVINCIAL GROSS DOMESTIC PRODUCT	177991	200840	222367	244498	271303	298197	332881
REAL PROVINCIAL GDP AT FACTOR COST	31500	32767	33414	33965	34673	35019	35920
DEFLATOR, PROV'L GDP AT FACTOR COST (71=1.0)	5.01	5.45	5.92	6.39	6.93	7.52	8.21
REAL PERSONAL DISP. INCOME PER CAP. ($'000)	3.836	3.914	3.928	3.945	4.028	4.053	4.184
ANNUAL WAGES & SALARIES PER EMPLOYEE ($'000)	32.408	34.816	37.962	41.739	45.559	50.268	55.208
GDP AT FACTOR COST PER EMPLOYEE ($71 '000)	10.987	11.141	11.301	11.406	11.450	11.441	11.619
LABOUR FORCE ('000)	3247	3297	3296	3317	3356	3385	3408
EMPLOYMENT ('000)	2867	2941	2957	2978	3028	3061	3091
UNEMPLOYMENT RATE (%)	11.70	10.79	10.30	10.23	9.76	9.57	9.30
REAL DOMESTIC PRODUCT - MANUFACTURING	7301	7550	7618	7654	7712	7649	7651
RDP - GOODS-PRODUCING INDUSTRIES	10658	11070	11190	11266	11382	11335	11401
RDP - SERVICE INDUSTRIES	17634	18455	18950	19393	19954	20318	21122

NOTE - NOMINAL VALUES IN MILLIONS OF CURRENT DOLLARS
- REAL VALUES IN MILLIONS OF 1971 DOLLARS

INSTITUTE FOR POLICY ANALYSIS, UNIVERSITY OF TORONTO
PRISM82A *** REVISED LO-TREND BASE CASE *** SEPT/82

Appendix.Table 15.5 (cont'd.)
SUMMARY - PROVINCIAL ECONOMY

(PERCENTAGE CHANGE; * INDICATES CHANGE IN LEVELS)

QUEBEC

	1981	1982	1983	1984	1985	1986	1987	1988
PROVINCIAL GROSS DOMESTIC PRODUCT	13.82	1.00	12.52	13.80	14.63	12.37	11.55	11.85
REAL PROVINCIAL GDP AT FACTOR COST	3.94	-9.31	2.90	3.70	3.95	2.90	1.81	2.70
DEFLATOR, PROV'L GDP AT FACTOR COST	7.19	11.60	9.83	9.90	10.75	9.25	9.32	9.03
REAL PERSONAL DISP. INCOME PER CAP.	1.60	-3.35	0.79	2.71	1.32	0.89	1.29	2.11
ANNUAL WAGES & SALARIES PER EMPLOYEE	12.61	7.44	6.05	9.34	6.73	9.26	10.90	9.58
GDP AT FACTOR COST PER EMPLOYEE	3.27	-5.02	2.59	1.70	0.60	0.75	0.32	1.59
LABOUR FORCE	1.27	-0.31	0.64	1.50	1.67	1.76	1.40	0.69
EMPLOYMENT	0.65	-4.52	0.30	1.96	3.34	2.13	1.48	1.10
UNEMPLOYMENT RATE (%) *	0.56	3.78	0.29	-0.39	-1.41	-0.31	-0.07	-0.36
REAL DOMESTIC PRODUCT - MANUFACTURING	3.49	-10.78	2.73	2.37	3.65	1.98	0.41	2.17
RDP - GOODS-PRODUCING INDUSTRIES	3.14	-13.91	3.90	3.84	4.04	1.45	0.03	2.01
RDP - SERVICE INDUSTRIES	5.01	-8.13	2.80	4.34	4.72	4.24	3.13	3.47

INSTITUTE FOR POLICY ANALYSIS, UNIVERSITY OF TORONTO
PRISM82A * REVISED LO-TREND BASE CASE *** SEPT/82**

Appendix Table 15.5 (cont'd.)
SUMMARY - PROVINCIAL ECONOMY

(PERCENTAGE CHANGE; * INDICATES CHANGE IN LEVELS)

QUEBEC

	1989	1990	1991	1992	1993	1994	1995
PROVINCIAL GROSS DOMESTIC PRODUCT	10.98	12.84	10.72	9.95	10.96	9.91	11.63
REAL PROVINCIAL GDP AT FACTOR COST	2.54	4.02	1.97	1.65	2.08	1.00	2.57
DEFLATOR, PROV'L GDP AT FACTOR COST	8.28	8.67	8.63	7.93	8.50	8.54	9.12
REAL PERSONAL DISP. INCOME PER CAP.	1.33	2.02	0.38	0.43	2.11	0.61	3.25
ANNUAL WAGES & SALARIES PER EMPLOYEE	9.09	7.43	9.04	9.95	9.15	10.34	9.83
GDP AT FACTOR COST PER EMPLOYEE	1.54	1.40	1.44	0.93	0.39	-0.07	1.56
LABOUR FORCE	0.72	1.54	-0.01	0.63	1.17	0.86	0.69
EMPLOYMENT	0.99	2.59	0.53	0.72	1.69	1.07	1.00
UNEMPLOYMENT RATE (%) *	-0.24	-0.91	-0.48	-0.07	-0.46	-0.19	-0.28
REAL DOMESTIC PRODUCT - MANUFACTURING	1.58	3.42	0.90	0.47	0.75	-0.81	0.03
RDP - GOODS-PRODUCING INDUSTRIES	1.77	3.86	1.09	0.68	1.03	-0.42	0.58
RDP - SERVICE INDUSTRIES	3.29	4.66	2.68	2.34	2.89	1.83	3.95

INSTITUTE FOR POLICY ANALYSIS, UNIVERSITY OF TORONTO
PRISM82A *** REVISED LO-TREND BASE CASE *** SEPT/82

Appendix Table 15.6
SUMMARY - PROVINCIAL ECONOMY

ONTARIO

	1981	1982	1983	1984	1985	1986	1987	1988
PROVINCIAL GROSS DOMESTIC PRODUCT	127275	137637	153369	174244	201305	226438	252831	283647
REAL PROVINCIAL GDP AT FACTOR COST	46118	44807	45646	47424	49736	51264	52270	53807
DEFLATOR, PROV'L GDP AT FACTOR COST (71=1.0)	2.39	2.67	2.93	3.21	3.55	3.88	4.24	4.63
REAL PERSONAL DISP. INCOME PER CAP. ($'000)	4.175	4.159	4.114	4.178	4.228	4.238	4.264	4.328
ANNUAL WAGES & SALARIES PER EMPLOYEE ($'000)	17.467	19.494	20.590	22.463	23.989	26.186	28.959	31.729
GDP AT FACTOR COST PER EMPLOYEE ($71 '000)	11.017	10.940	11.181	11.363	11.412	11.482	11.495	11.656
LABOUR FORCE ('000)	4482	4516	4589	4692	4784	4877	4966	5023
EMPLOYMENT ('000)	4186	4096	4083	4174	4358	4465	4547	4616
UNEMPLOYMENT RATE (%)	6.60	9.30	11.03	11.06	8.90	8.45	8.43	8.10
REAL DOMESTIC PRODUCT - MANUFACTURING	13844	12816	12885	13411	13976	14266	14273	14560
RDP - GOODS-PRODUCING INDUSTRIES	18528	17524	17524	17985	18787	19102	19077	19456
RDP - SERVICE INDUSTRIES	22656	22283	23058	24285	25712	26865	27828	28910

NOTE - NOMINAL VALUES IN MILLIONS OF CURRENT DOLLARS
     - REAL VALUES IN MILLIONS OF 1971 DOLLARS

INSTITUTE FOR POLICY ANALYSIS, UNIVERSITY OF TORONTO
PRISM82A * REVISED LO-TREND BASE CASE *** SEPT/82**

Appendix Table 15.6 (cont'd.)

SUMMARY - PROVINCIAL ECONOMY

ONTARIO

	1989	1990	1991	1992	1993	1994	1995
PROVINCIAL GROSS DOMESTIC PRODUCT	315603	357169	397511	437154	486082	534936	598303
REAL PROVINCIAL GDP AT FACTOR COST	55239	57594	58937	60008	61355	62053	63676
DEFLATOR, PROV'L GDP AT FACTOR COST (71=1.0)	5.01	5.45	5.91	6.39	6.93	7.53	8.22
REAL PERSONAL DISP. INCOME PER CAP. ($/000)	4.357	4.422	4.422	4.418	4.482	4.487	4.609
ANNUAL WAGES & SALARIES PER EMPLOYEE ($/000)	34.601	37.210	40.602	44.559	48.546	53.418	58.747
GDP AT FACTOR COST PER EMPLOYEE ($71/000)	11.812	11.958	12.119	12.211	12.238	12.212	12.382
LABOUR FORCE ('000)	5080	5162	5199	5263	5339	5409	5475
EMPLOYMENT ('000)	4676	4816	4863	4914	5013	5081	5143
UNEMPLOYMENT RATE (%)	7.95	6.69	6.45	6.62	6.10	6.05	6.07
REAL DOMESTIC PRODUCT - MANUFACTURING	14758	15299	15538	15611	15742	15626	15606
RDP - GOODS-PRODUCING INDUSTRIES	19777	20583	20917	21069	21302	21222	21314
RDP - SERVICE INDUSTRIES	29946	31419	32352	33197	34238	34945	36405

NOTE - NOMINAL VALUES IN MILLIONS OF CURRENT DOLLARS
       - REAL VALUES IN MILLIONS OF 1971 DOLLARS

INSTITUTE FOR POLICY ANALYSIS, UNIVERSITY OF TORONTO
PRISM82A *** REVISED LO-TREND BASE CASE *** SEPT/82

Appendix Table 15.6 (cont'd.)
SUMMARY - PROVINCIAL ECONOMY

(PERCENTAGE CHANGE; * INDICATES CHANGE IN LEVELS)

ONTARIO

	1981	1982	1983	1984	1985	1986	1987	1988
PROVINCIAL GROSS DOMESTIC PRODUCT	13.90	8.14	11.43	13.61	15.53	12.49	11.66	12.19
REAL PROVINCIAL GDP AT FACTOR COST	4.06	-2.84	1.87	3.89	4.88	3.07	1.96	2.94
DEFLATOR, PROV'L GDP AT FACTOR COST	7.59	11.65	9.82	9.56	10.63	9.26	9.38	9.09
REAL PERSONAL DISP. INCOME PER CAP.	1.76	-0.38	-1.08	1.56	1.20	0.22	0.63	1.50
ANNUAL WAGES & SALARIES PER EMPLOYEE	10.11	11.61	5.62	9.09	6.80	9.16	10.59	9.57
GDP AT FACTOR COST PER EMPLOYEE	1.10	-0.70	2.20	1.63	0.44	0.61	0.12	1.40
LABOUR FORCE	2.64	0.76	1.61	2.26	1.95	1.95	1.82	1.15
EMPLOYMENT	2.93	-2.16	-0.32	2.23	4.42	2.45	1.84	1.52
UNEMPLOYMENT RATE (%) *	-0.27	2.70	1.73	0.03	-2.16	-0.45	-0.02	-0.34
REAL DOMESTIC PRODUCT - MANUFACTURING	3.77	-7.42	0.54	4.08	4.21	2.08	0.05	2.01
RDP - GOODS-PRODUCING INDUSTRIES	4.90	-5.42	0.00	2.63	4.46	1.67	-0.13	1.99
RDP - SERVICE INDUSTRIES	4.21	-1.64	3.48	5.32	5.88	4.49	3.58	3.89

INSTITUTE FOR POLICY ANALYSIS, UNIVERSITY OF TORONTO
PRISM82A * REVISED LO-TREND BASE CASE *** SEPT/82**

Appendix Table 15.6 (cont'd.)

SUMMARY - PROVINCIAL ECONOMY

(PERCENTAGE CHANGE; * INDICATES CHANGE IN LEVELS)

ONTARIO

	1989	1990	1991	1992	1993	1994	1995
PROVINCIAL GROSS DOMESTIC PRODUCT	11.27	13.17	11.30	9.97	11.19	10.05	11.85
REAL PROVINCIAL GDP AT FACTOR COST	2.66	4.26	2.33	1.82	2.24	1.14	2.61
DEFLATOR, PROV'L GDP AT FACTOR COST	8.33	8.64	8.57	8.00	8.52	8.59	9.20
REAL PERSONAL DISP. INCOME PER CAP.	0.66	1.49	-0.00	-0.08	1.44	0.12	2.72
ANNUAL WAGES & SALARIES PER EMPLOYEE	9.05	7.54	9.12	9.75	8.95	10.04	9.98
GDP AT FACTOR COST PER EMPLOYEE	1.34	1.23	1.34	0.76	0.22	-0.22	1.40
LABOUR FORCE	1.14	1.61	0.71	1.23	1.45	1.31	1.22
EMPLOYMENT	1.30	2.99	0.98	1.05	2.02	1.36	1.20
UNEMPLOYMENT RATE (%) *	-0.15	-1.25	-0.24	0.17	-0.52	-0.05	0.01
REAL DOMESTIC PRODUCT - MANUFACTURING	1.36	3.66	1.56	0.47	0.84	-0.74	-0.13
RDP - GOODS-PRODUCING INDUSTRIES	1.65	4.08	1.62	0.73	1.10	-0.38	0.43
RDP - SERVICE INDUSTRIES	3.58	4.92	2.97	2.61	3.13	2.07	4.18

INSTITUTE FOR POLICY ANALYSIS, UNIVERSITY OF TORONTO
PRISM82A *** REVISED LO-TREND BASE CASE *** SEPT/82

Appendix Table 15.7
SUMMARY - PROVINCIAL ECONOMY

MANITOBA

	1981	1982	1983	1984	1985	1986	1987	1988
PROVINCIAL GROSS DOMESTIC PRODUCT	12574	13033	14595	16572	19211	21633	24184	27118
REAL PROVINCIAL GDP AT FACTOR COST	4552	4274	4392	4554	4757	4908	5013	5160
DEFLATOR, PROV'L GDP AT FACTOR COST (71=1.0)	2.46	2.73	3.00	3.29	3.66	4.00	4.38	4.79
REAL PERSONAL DISP. INCOME PER CAP. ($'000)	3.683	3.630	3.643	3.722	3.780	3.812	3.860	3.933
ANNUAL WAGES & SALARIES PER EMPLOYEE ($'000)	15.080	16.166	17.155	18.848	20.299	22.306	24.865	27.397
GDP AT FACTOR COST PER EMPLOYEE ($71 '000)	9.846	9.377	9.628	9.814	9.901	10.005	10.059	10.239
LABOUR FORCE ('000)	492	494	497	505	513	525	533	537
EMPLOYMENT ('000)	462	456	456	464	480	491	498	504
UNEMPLOYMENT RATE (%)	6.08	7.76	8.19	8.08	6.37	6.50	6.53	6.19
REAL DOMESTIC PRODUCT - MANUFACTURING	659	611	625	646	676	699	711	736
RDP - GOODS-PRODUCING INDUSTRIES	1371	1257	1286	1321	1383	1408	1417	1453
RDP - SERVICE INDUSTRIES	2622	2454	2535	2656	2793	2913	3003	3108

NOTE - NOMINAL VALUES IN MILLIONS OF CURRENT DOLLARS
     - REAL VALUES IN MILLIONS OF 1971 DOLLARS

INSTITUTE FOR POLICY ANALYSIS, UNIVERSITY OF TORONTO
PRISM82A *** REVISED LO-TREND BASE CASE *** SEPT/82

Appendix Table 15.7 (cont'd.)

MANITOBA
SUMMARY - PROVINCIAL ECONOMY

	1989	1990	1991	1992	1993	1994	1995
PROVINCIAL GROSS DOMESTIC PRODUCT	30237	34345	38130	41859	46590	51280	57239
REAL PROVINCIAL GDP AT FACTOR COST	5305	5531	5656	5769	5909	5989	6156
DEFLATOR, PROV'L GDP AT FACTOR COST (71=1.0)	5.19	5.66	6.15	6.64	7.22	7.84	8.55
REAL PERSONAL DISP. INCOME PER CAP. ($'000)	3.982	4.063	4.077	4.097	4.175	4.204	4.330
ANNUAL WAGES & SALARIES PER EMPLOYEE ($'000)	30.069	32.576	35.758	39.462	43.271	47.941	52.925
GDP AT FACTOR COST PER EMPLOYEE ($71'000)	10.419	10.593	10.774	10.903	10.973	10.995	11.188
LABOUR FORCE ('000)	544	553	556	564	572	578	584
EMPLOYMENT ('000)	509	522	525	529	539	545	550
UNEMPLOYMENT RATE (%)	6.37	5.62	5.56	6.11	5.78	5.78	5.72
REAL DOMESTIC PRODUCT - MANUFACTURING	756	792	810	824	839	840	849
RDP - GOODS-PRODUCING INDUSTRIES	1487	1554	1581	1603	1632	1636	1656
RDP - SERVICE INDUSTRIES	3211	3363	3454	3537	3643	3711	3851

NOTE - NOMINAL VALUES IN MILLIONS OF CURRENT DOLLARS
        - REAL VALUES IN MILLIONS OF 1971 DOLLARS

INSTITUTE FOR POLICY ANALYSIS, UNIVERSITY OF TORONTO
PRISM82A *** REVISED LO-TREND BASE CASE *** SEPT/82

Appendix Table 15.7 (cont'd.)
SUMMARY - PROVINCIAL ECONOMY

(PERCENTAGE CHANGE; * INDICATES CHANGE IN LEVELS)

MANITOBA

	1981	1982	1983	1984	1985	1986	1987	1988
PROVINCIAL GROSS DOMESTIC PRODUCT	13.23	3.65	11.98	13.55	15.92	12.61	11.79	12.13
REAL PROVINCIAL GDP AT FACTOR COST	4.41	-6.11	2.75	3.70	4.45	3.17	2.14	2.95
DEFLATOR, PROV'L GDP AT FACTOR COST	7.05	11.22	9.77	9.87	11.07	9.37	9.59	9.18
REAL PERSONAL DISP. INCOME PER CAP.	2.87	-1.43	0.34	2.17	1.55	0.85	1.26	1.88
ANNUAL WAGES & SALARIES PER EMPLOYEE	13.13	7.20	6.12	9.87	7.70	9.89	11.47	10.18
GDP AT FACTOR COST PER EMPLOYEE	3.65	-4.76	2.68	1.93	0.89	1.05	0.54	1.79
LABOUR FORCE	1.35	0.39	0.52	1.62	1.64	2.24	1.62	0.77
EMPLOYMENT	0.74	-1.41	0.06	1.74	3.53	2.10	1.59	1.14
UNEMPLOYMENT RATE (%) *	0.57	1.69	0.42	-0.10	-1.71	0.14	0.03	-0.34
REAL DOMESTIC PRODUCT - MANUFACTURING	5.96	-7.29	2.36	3.24	4.72	3.40	1.68	3.49
RDP - GOODS-PRODUCING INDUSTRIES	7.36	-8.29	2.32	2.71	4.66	1.85	0.61	2.56
RDP - SERVICE INDUSTRIES	3.79	-6.41	3.33	4.78	5.15	4.28	3.11	3.49

INSTITUTE FOR POLICY ANALYSIS, UNIVERSITY OF TORONTO
PRISM82A * REVISED LO-TREND BASE CASE *** SEPT/82**

Appendix Table 15.7 (cont'd.)
SUMMARY - PROVINCIAL ECONOMY

(PERCENTAGE CHANGE; * INDICATES CHANGE IN LEVELS)

MANITOBA

	1989	1990	1991	1992	1993	1994	1995
PROVINCIAL GROSS DOMESTIC PRODUCT	11.50	13.59	11.02	9.78	11.30	10.07	11.62
REAL PROVINCIAL GDP AT FACTOR COST	2.81	4.26	2.26	1.99	2.43	1.35	2.78
DEFLATOR, PROV'L GDP AT FACTOR COST	8.47	8.94	8.76	7.98	8.62	8.65	9.08
REAL PERSONAL DISP. INCOME PER CAP.	1.26	2.02	0.36	0.49	1.89	0.71	2.99
ANNUAL WAGES & SALARIES PER EMPLOYEE	9.75	8.34	9.77	10.36	9.65	10.79	10.40
GDP AT FACTOR COST PER EMPLOYEE	1.76	1.67	1.70	1.20	0.65	0.20	1.75
LABOUR FORCE	1.23	1.73	0.48	1.37	1.42	1.15	0.95
EMPLOYMENT	1.03	2.54	0.54	0.79	1.78	1.15	1.02
UNEMPLOYMENT RATE (%) *	0.18	-0.75	-0.06	0.55	-0.33	0.00	-0.06
REAL DOMESTIC PRODUCT - MANUFACTURING	2.68	4.79	2.32	1.70	1.88	0.11	0.99
RDP - GOODS-PRODUCING INDUSTRIES	2.34	4.48	1.74	1.41	1.77	0.26	1.22
RDP - SERVICE INDUSTRIES	3.33	4.74	2.70	2.41	2.97	1.89	3.76

INSTITUTE FOR POLICY ANALYSIS, UNIVERSITY OF TORONTO
PRISM82A *** REVISED LO-TREND BASE CASE *** SEPT/82

Appendix Table 15.8
SUMMARY - PROVINCIAL ECONOMY

SASKATCHEWAN

	1981	1982	1983	1984	1985	1986	1987	1988
PROVINCIAL GROSS DOMESTIC PRODUCT	13973	14681	16413	18523	21304	24069	27073	30630
REAL PROVINCIAL GDP AT FACTOR COST	4239	4059	4193	4349	4566	4718	4836	5038
DEFLATOR, PROV'L GDP AT FACTOR COST (71=1.0)	2.96	3.31	3.62	3.95	4.38	4.79	5.27	5.74
REAL PERSONAL DISP. INCOME PER CAP. ($'000)	3.790	3.693	3.729	3.812	3.901	3.984	4.056	4.190
ANNUAL WAGES & SALARIES PER EMPLOYEE ($'000)	13.254	14.603	15.619	17.384	19.034	21.178	23.851	26.626
GDP AT FACTOR COST PER EMPLOYEE ($71 '000)	9.815	9.535	9.788	9.974	10.064	10.172	10.233	10.411
LABOUR FORCE ('000)	453	453	457	466	480	491	501	510
EMPLOYMENT ('000)	432	426	428	436	454	464	473	484
UNEMPLOYMENT RATE (%)	4.63	5.96	6.27	6.48	5.55	5.45	5.64	5.18
REAL DOMESTIC PRODUCT - MANUFACTURING	260	244	252	260	274	283	289	300
RDP - GOODS-PRODUCING INDUSTRIES	1612	1532	1560	1588	1654	1672	1680	1752
RDP - SERVICE INDUSTRIES	2121	2012	2106	2224	2365	2488	2588	2706

NOTE - NOMINAL VALUES IN MILLIONS OF CURRENT DOLLARS
       - REAL VALUES IN MILLIONS OF 1971 DOLLARS

INSTITUTE FOR POLICY ANALYSIS, UNIVERSITY OF TORONTO
PRISM82A *** REVISED LO-TREND BASE CASE *** SEPT/82

Appendix Table 15.8 (cont'd.)

SUMMARY - PROVINCIAL ECONOMY
SASKATCHEWAN

	1989	1990	1991	1992	1993	1994	1995
PROVINCIAL GROSS DOMESTIC PRODUCT	34319	39349	43899	48614	54926	60986	68673
REAL PROVINCIAL GDP AT FACTOR COST	5213	5480	5641	5799	5977	6112	6337
DEFLATOR, PROV'L GDP AT FACTOR COST (71=1.0)	6.22	6.78	7.37	7.97	8.74	9.51	10.38
REAL PERSONAL DISP. INCOME PER CAP. ($'000)	4.289	4.437	4.505	4.569	4.686	4.767	4.970
ANNUAL WAGES & SALARIES PER EMPLOYEE ($'000)	29.505	32.411	35.975	39.952	44.022	48.986	54.605
GDP AT FACTOR COST PER EMPLOYEE ($71,000)	10.586	10.754	10.928	11.059	11.146	11.177	11.385
LABOUR FORCE ('000)	519	534	539	548	559	570	579
EMPLOYMENT ('000)	492	510	516	524	536	547	557
UNEMPLOYMENT RATE (%)	5.07	4.55	4.14	4.26	4.13	4.05	3.89
REAL DOMESTIC PRODUCT - MANUFACTURING	310	327	336	344	353	356	362
RDP - GOODS-PRODUCING INDUSTRIES	1794	1880	1911	1945	1976	1996	2035
RDP - SERVICE INDUSTRIES	2825	2991	3108	3217	3348	3449	3620

NOTE - NOMINAL VALUES IN MILLIONS OF CURRENT DOLLARS
       - REAL VALUES IN MILLIONS OF 1971 DOLLARS

INSTITUTE FOR POLICY ANALYSIS, UNIVERSITY OF TORONTO
PRISM82A *** REVISED LO-TREND BASE CASE *** SEPT/82

Appendix Table 15.8 (cont'd.)
SUMMARY - PROVINCIAL ECONOMY

(PERCENTAGE CHANGE; * INDICATES CHANGE IN LEVELS)

SASKATCHEWAN

	1981	1982	1983	1984	1985	1986	1987	1988
PROVINCIAL GROSS DOMESTIC PRODUCT	7.14	5.07	11.80	12.86	15.01	12.98	12.48	13.14
REAL PROVINCIAL GDP AT FACTOR COST	2.48	-4.25	3.31	3.73	4.98	3.32	2.51	4.18
DEFLATOR, PROV'L GDP AT FACTOR COST	5.02	11.87	9.35	9.28	10.64	9.57	9.96	8.88
REAL PERSONAL DISP. INCOME PER CAP.	4.52	-2.55	0.97	2.23	2.32	2.15	1.81	3.30
ANNUAL WAGES & SALARIES PER EMPLOYEE	10.66	10.18	6.95	11.30	9.49	11.27	12.62	11.64
GDP AT FACTOR COST PER EMPLOYEE	0.44	-2.86	2.66	1.91	0.90	1.07	0.60	1.74
LABOUR FORCE	2.27	-0.03	0.97	2.01	3.02	2.13	2.10	1.90
EMPLOYMENT	2.04	-1.43	0.64	1.79	4.04	2.23	1.90	2.40
UNEMPLOYMENT RATE (%) *	0.21	1.33	0.31	0.21	-0.92	-0.10	0.18	-0.46
REAL DOMESTIC PRODUCT - MANUFACTURING	4.45	-6.19	3.26	3.39	5.11	3.42	2.13	3.97
RDP - GOODS-PRODUCING INDUSTRIES	0.53	-4.94	1.78	1.81	4.14	1.11	0.49	4.25
RDP - SERVICE INDUSTRIES	4.35	-5.15	4.66	5.62	6.31	5.20	4.02	4.57

INSTITUTE FOR POLICY ANALYSIS, UNIVERSITY OF TORONTO
PRISM82A * REVISED LO-TREND BASE CASE *** SEPT/82**

Appendix Table 15.8 (cont'd.)
SUMMARY - PROVINCIAL ECONOMY

(PERCENTAGE CHANGE;   * INDICATES CHANGE IN LEVELS)

SASKATCHEWAN

	1989	1990	1991	1992	1993	1994	1995
PROVINCIAL GROSS DOMESTIC PRODUCT	12.04	14.66	11.56	10.74	12.99	11.03	12.60
REAL PROVINCIAL GDP AT FACTOR COST	3.47	5.11	2.95	2.80	3.05	2.26	3.69
DEFLATOR, PROV'L GDP AT FACTOR COST	8.33	9.04	8.70	8.16	9.69	8.76	9.11
REAL PERSONAL DISP. INCOME PER CAP.	2.35	3.45	1.54	1.43	2.55	1.73	4.25
ANNUAL WAGES & SALARIES PER EMPLOYEE	10.81	9.87	10.98	11.06	10.19	11.28	11.47
GDP AT FACTOR COST PER EMPLOYEE	1.68	1.59	1.62	1.19	0.78	0.28	1.87
LABOUR FORCE	1.64	2.91	0.87	1.71	2.12	1.89	1.62
EMPLOYMENT	1.76	3.47	1.31	1.59	2.25	1.97	1.79
UNEMPLOYMENT RATE (%) *	-0.11	-0.51	-0.41	0.12	-0.12	-0.08	-0.17
REAL DOMESTIC PRODUCT - MANUFACTURING	3.34	5.36	2.71	2.34	2.62	0.91	1.77
RDP - GOODS-PRODUCING INDUSTRIES	2.42	4.78	1.65	1.79	1.61	0.97	1.99
RDP - SERVICE INDUSTRIES	4.40	5.90	3.88	3.52	4.08	3.02	4.94

INSTITUTE FOR POLICY ANALYSIS, UNIVERSITY OF TORONTO
PRISM82A *** REVISED LO-TREND BASE CASE *** SEPT/82

Appendix Table 15.9
SUMMARY - PROVINCIAL ECONOMY

ALBERTA

	1981	1982	1983	1984	1985	1986	1987	1988
PROVINCIAL GROSS DOMESTIC PRODUCT	46572	51192	58471	66333	77103	88289	99516	111918
REAL PROVINCIAL GDP AT FACTOR COST	14275	13798	14308	14887	15761	16384	16833	17421
DEFLATOR, PROV'L GDP AT FACTOR COST (71=1.0)	3.06	3.49	3.86	4.22	4.64	5.12	5.62	6.11
REAL PERSONAL DISP. INCOME PER CAP. ($'000)	4.387	4.294	4.302	4.410	4.542	4.629	4.682	4.786
ANNUAL WAGES & SALARIES PER EMPLOYEE ($'000)	17.157	19.441	20.962	23.393	25.615	28.622	32.322	35.990
GDP AT FACTOR COST PER EMPLOYEE ($71 '000)	13.054	12.959	13.290	13.485	13.511	13.635	13.705	13.914
LABOUR FORCE ('000)	1137	1146	1168	1200	1243	1277	1313	1338
EMPLOYMENT ('000)	1093	1065	1077	1104	1167	1202	1228	1252
UNEMPLOYMENT RATE (%)	3.82	7.11	7.81	8.02	6.17	5.93	6.47	6.43
REAL DOMESTIC PRODUCT - MANUFACTURING	1246	1179	1229	1287	1371	1429	1468	1537
RDP - GOODS-PRODUCING INDUSTRIES	5564	5217	5326	5433	5765	5912	5963	6098
RDP - SERVICE INDUSTRIES	7271	7074	7423	7859	8365	8806	9170	9584

NOTE - NOMINAL VALUES IN MILLIONS OF CURRENT DOLLARS
       - REAL VALUES IN MILLIONS OF 1971 DOLLARS

INSTITUTE FOR POLICY ANALYSIS, UNIVERSITY OF TORONTO
PRISM82A *** REVISED LO-TREND BASE CASE *** SEPT/82

Appendix Table 15.9 (cont'd.)
SUMMARY - PROVINCIAL ECONOMY

ALBERTA

	1989	1990	1991	1992	1993	1994	1995
PROVINCIAL GROSS DOMESTIC PRODUCT	125483	144674	162265	182357	207849	232196	264383
REAL PROVINCIAL GDP AT FACTOR COST	18064	19114	19752	20429	21270	21860	22818
DEFLATOR, PROV'L GDP AT FACTOR COST (71=1.0)	6.61	7.20	7.83	8.52	9.33	10.14	11.09
REAL PERSONAL DISP. INCOME PER CAP. ($'000)	4.848	5.013	5.067	5.128	5.275	5.334	5.565
ANNUAL WAGES & SALARIES PER EMPLOYEE ($'000)	39.890	43.749	48.458	53.820	59.390	65.909	73.412
GDP AT FACTOR COST PER EMPLOYEE ($71 '000)	14.118	14.283	14.483	14.635	14.706	14.722	14.975
LABOUR FORCE ('000)	1370	1416	1438	1471	1513	1548	1582
EMPLOYMENT ('000)	1280	1338	1364	1396	1446	1485	1524
UNEMPLOYMENT RATE (%)	6.64	5.51	5.16	5.12	4.42	4.08	3.65
REAL DOMESTIC PRODUCT - MANUFACTURING	1602	1706	1768	1824	1886	1917	1966
RDP - GOODS-PRODUCING INDUSTRIES	6294	6728	6923	7162	7472	7639	7905
RDP - SERVICE INDUSTRIES	9994	10569	10968	11360	11840	12211	12848

NOTE - NOMINAL VALUES IN MILLIONS OF CURRENT DOLLARS
      - REAL VALUES IN MILLIONS OF 1971 DOLLARS

INSTITUTE FOR POLICY ANALYSIS, UNIVERSITY OF TORONTO
PRISM82A *** REVISED LO-TREND BASE CASE *** SEPT/82

Appendix Table 15.9 (cont'd.)
SUMMARY - PROVINCIAL ECONOMY

(PERCENTAGE CHANGE; * INDICATES CHANGE IN LEVELS)

ALBERTA

	1981	1982	1983	1984	1985	1986	1987	1988
PROVINCIAL GROSS DOMESTIC PRODUCT	15.93	9.92	14.22	13.45	16.24	14.51	12.72	12.46
REAL PROVINCIAL GDP AT FACTOR COST	4.58	-3.34	3.69	4.05	5.87	3.95	2.74	3.49
DEFLATOR, PROV'L GDP AT FACTOR COST	10.25	14.33	10.62	9.24	9.94	10.28	9.79	8.81
REAL PERSONAL DISP. INCOME PER CAP.	3.11	-2.13	0.20	2.51	2.99	1.93	1.14	2.22
ANNUAL WAGES & SALARIES PER EMPLOYEE	11.39	13.31	7.82	11.60	9.50	11.74	12.92	11.35
GDP AT FACTOR COST PER EMPLOYEE	-1.31	-0.73	2.55	1.47	0.19	0.92	0.51	1.52
LABOUR FORCE	6.10	0.82	1.87	2.79	3.59	2.73	2.81	1.88
EMPLOYMENT	5.97	-2.63	1.12	2.55	5.67	3.00	2.22	1.94
UNEMPLOYMENT RATE (%) *	0.12	3.29	0.69	0.22	-1.85	-0.24	0.54	-0.05
REAL DOMESTIC PRODUCT - MANUFACTURING	-0.61	-5.31	4.17	4.78	6.53	4.19	2.70	4.70
RDP - GOODS-PRODUCING INDUSTRIES	1.88	-6.24	2.09	2.01	6.11	2.56	0.87	2.26
RDP - SERVICE INDUSTRIES	6.70	-2.71	4.94	5.87	6.44	5.28	4.13	4.52

INSTITUTE FOR POLICY ANALYSIS, UNIVERSITY OF TORONTO
PRISM82A *** REVISED LO-TREND BASE CASE *** SEPT/82
Appendix Table 15.9 (cont'd.)

SUMMARY - PROVINCIAL ECONOMY
(PERCENTAGE CHANGE; * INDICATES CHANGE IN LEVELS)

ALBERTA

	1989	1990	1991	1992	1993	1994	1995
PROVINCIAL GROSS DOMESTIC PRODUCT	12.12	15.29	12.16	12.38	13.98	11.71	13.86
REAL PROVINCIAL GDP AT FACTOR COST	3.69	5.81	3.33	3.43	4.12	2.77	4.38
DEFLATOR, PROV'L GDP AT FACTOR COST	8.13	8.96	8.65	8.85	9.45	8.74	9.35
REAL PERSONAL DISP. INCOME PER CAP.	1.29	3.41	1.06	1.22	2.85	1.12	4.34
ANNUAL WAGES & SALARIES PER EMPLOYEE	10.84	9.67	10.76	11.06	10.35	10.98	11.38
GDP AT FACTOR COST PER EMPLOYEE	1.47	1.17	1.40	1.05	0.49	0.11	1.72
LABOUR FORCE	2.43	3.34	1.54	2.31	2.85	2.30	2.17
EMPLOYMENT	2.20	4.59	1.91	2.35	3.61	2.66	2.62
UNEMPLOYMENT RATE (%) *	0.21	-1.13	-0.34	-0.04	-0.71	-0.34	-0.43
REAL DOMESTIC PRODUCT - MANUFACTURING	4.28	6.49	3.60	3.18	3.41	1.66	2.54
RDP - GOODS-PRODUCING INDUSTRIES	3.21	6.88	2.90	3.46	4.33	2.23	3.49
RDP - SERVICE INDUSTRIES	4.27	5.76	3.78	3.57	4.23	3.14	5.21

INSTITUTE FOR POLICY ANALYSIS, UNIVERSITY OF TORONTO
PRISM82A *** REVISED LO-TREND BASE CASE *** SEPT/82

Appendix Table 15.10

SUMMARY - PROVINCIAL ECONOMY

BRITISH COLUMBIA

	1981	1982	1983	1984	1985	1986	1987	1988
PROVINCIAL GROSS DOMESTIC PRODUCT	41181	43879	49803	58437	68672	78301	88474	100223
REAL PROVINCIAL GDP AT FACTOR COST	14636	14022	14544	15395	16325	16974	17473	18143
DEFLATOR, PROV'L GDP AT FACTOR COST (71=1.0)	2.49	2.79	3.06	3.40	3.78	4.15	4.56	4.98
REAL PERSONAL DISP. INCOME PER CAP. ($'000)	4.250	4.176	4.182	4.310	4.406	4.434	4.477	4.556
ANNUAL WAGES & SALARIES PER EMPLOYEE ($'000)	18.180	20.700	22.066	24.374	26.376	29.088	32.456	35.769
GDP AT FACTOR COST PER EMPLOYEE ($71 '000)	11.734	11.868	12.155	12.369	12.420	12.517	12.563	12.752
LABOUR FORCE ('000)	1337	1343	1369	1411	1452	1498	1538	1569
EMPLOYMENT ('000)	1247	1181	1196	1245	1314	1356	1391	1423
UNEMPLOYMENT RATE (%)	6.72	12.02	12.57	11.77	9.48	9.45	9.58	9.35
REAL DOMESTIC PRODUCT - MANUFACTURING	2287	2130	2187	2249	2367	2438	2482	2567
RDP - GOODS-PRODUCING INDUSTRIES	4717	4389	4536	4857	5179	5302	5360	5548
RDP - SERVICE INDUSTRIES	8259	7930	8263	8755	9323	9808	10206	10646

NOTE - NOMINAL VALUES IN MILLIONS OF CURRENT DOLLARS
       - REAL VALUES IN MILLIONS OF 1971 DOLLARS

INSTITUTE FOR POLICY ANALYSIS, UNIVERSITY OF TORONTO
PRISM82A *** REVISED LO-TREND BASE CASE *** SEPT/82

Appendix Table 15.10 (cont'd.)

SUMMARY - PROVINCIAL ECONOMY

BRITISH COLUMBIA

	1989	1990	1991	1992	1993	1994	1995
PROVINCIAL GROSS DOMESTIC PRODUCT	112822	129626	146026	163050	181723	202794	229747
REAL PROVINCIAL GDP AT FACTOR COST	18824	19848	20498	21130	21807	22335	23201
DEFLATOR, PROV'L GDP AT FACTOR COST (71=1.0)	5.40	5.89	6.42	6.96	7.51	8.17	8.94
REAL PERSONAL DISP. INCOME PER CAP. ($'000)	4.604	4.705	4.718	4.739	4.826	4.856	5.012
ANNUAL WAGES & SALARIES PER EMPLOYEE ($'000)	39.252	42.559	46.734	51.524	56.384	62.289	68.786
GDP AT FACTOR COST PER EMPLOYEE ($71 '000)	12.938	13.098	13.277	13.400	13.431	13.428	13.640
LABOUR FORCE ('000)	1605	1652	1678	1718	1760	1799	1836
EMPLOYMENT ('000)	1455	1515	1544	1577	1624	1663	1701
UNEMPLOYMENT RATE (%)	9.36	8.25	8.01	8.21	7.74	7.56	7.37
REAL DOMESTIC PRODUCT - MANUFACTURING	2649	2788	2856	2923	3000	3034	3083
RDP - GOODS-PRODUCING INDUSTRIES	5757	6129	6328	6528	6693	6832	7033
RDP - SERVICE INDUSTRIES	11077	11685	12092	12478	12943	13287	13904

NOTE - NOMINAL VALUES IN MILLIONS OF CURRENT DOLLARS
       - REAL VALUES IN MILLIONS OF 1971 DOLLARS

INSTITUTE FOR POLICY ANALYSIS, UNIVERSITY OF TORONTO
PRISM82A *** REVISED LO-TREND BASE CASE *** SEPT/82

Appendix Table 15.10 (cont'd.)
SUMMARY - PROVINCIAL ECONOMY

(PERCENTAGE CHANGE; * INDICATES CHANGE IN LEVELS)

BRITISH COLUMBIA

	1981	1982	1983	1984	1985	1986	1987	1988
PROVINCIAL GROSS DOMESTIC PRODUCT	12.80	6.55	13.50	17.34	17.52	14.02	12.99	13.28
REAL PROVINCIAL GDP AT FACTOR COST	3.72	-4.20	3.72	5.85	6.04	3.98	2.94	3.83
DEFLATOR, PROV'L GDP AT FACTOR COST	7.23	11.70	9.94	11.10	11.11	9.81	9.74	9.25
REAL PERSONAL DISP. INCOME PER CAP.	-0.61	-1.74	0.14	3.07	2.23	0.63	0.97	1.77
ANNUAL WAGES & SALARIES PER EMPLOYEE	7.21	13.86	6.60	10.46	8.22	10.28	11.58	10.21
GDP AT FACTOR COST PER EMPLOYEE	-0.95	1.14	2.42	1.76	0.41	0.78	0.37	1.51
LABOUR FORCE	4.66	0.43	1.91	3.07	2.94	3.14	2.71	2.03
EMPLOYMENT	4.71	-5.28	1.27	4.02	5.61	3.17	2.56	2.29
UNEMPLOYMENT RATE (%) *	-0.04	5.30	0.55	-0.81	-2.29	-0.03	0.13	-0.23
REAL DOMESTIC PRODUCT - MANUFACTURING	0.18	-6.83	2.64	2.84	5.28	3.00	1.80	3.39
RDP - GOODS-PRODUCING INDUSTRIES	1.47	-6.96	3.35	7.08	6.64	2.37	1.10	3.51
RDP - SERVICE INDUSTRIES	5.29	-3.99	4.20	5.95	6.48	5.20	4.06	4.30

INSTITUTE FOR POLICY ANALYSIS, UNIVERSITY OF TORONTO
PRISM32A *** REVISED LO-TREND BASE CASE *** SEPT/82
Appendix Table 15.10 (cont'd.)

SUMMARY - PROVINCIAL ECONOMY
(PERCENTAGE CHANGE; * INDICATES CHANGE IN LEVELS)

BRITISH COLUMBIA

	1989	1990	1991	1992	1993	1994	1995
PROVINCIAL GROSS DOMESTIC PRODUCT	12.57	14.89	12.65	11.66	11.45	11.60	13.29
REAL PROVINCIAL GDP AT FACTOR COST	3.76	5.44	3.28	3.08	3.20	2.42	3.88
DEFLATOR, PROV'L GDP AT FACTOR COST	8.47	9.02	9.04	8.43	7.86	8.86	9.34
REAL PERSONAL DISP. INCOME PER CAP.	1.06	2.19	0.27	0.44	1.85	0.61	3.22
ANNUAL WAGES & SALARIES PER EMPLOYEE	9.74	8.42	9.81	10.25	9.43	10.47	10.43
GDP AT FACTOR COST PER EMPLOYEE	1.46	1.23	1.37	0.92	0.23	-0.02	1.58
LABOUR FORCE	2.28	2.89	1.62	2.36	2.44	2.25	2.05
EMPLOYMENT	2.26	4.15	1.88	2.14	2.96	2.44	2.27
UNEMPLOYMENT RATE (%) *	0.02	-1.11	-0.24	0.19	-0.47	-0.18	-0.19
REAL DOMESTIC PRODUCT - MANUFACTURING	3.21	5.27	2.44	2.34	2.62	1.15	1.61
RDP - GOODS-PRODUCING INDUSTRIES	3.76	6.46	3.25	3.17	2.53	2.07	2.95
RDP - SERVICE INDUSTRIES	4.05	5.49	3.48	3.19	3.73	2.66	4.64

INSTITUTE FOR POLICY ANALYSIS, UNIVERSITY OF TORONTO
PRISM82A *** REVISED LO-TREND BASE CASE *** SEPT/82

Appendix Table 16.1
REAL DOMESTIC PRODUCT BY INDUSTRY
NOVA SCOTIA

(MILLIONS OF 1971 DOLLARS)

	1981	1982	1983	1984	1985	1986	1987	1988
1 AGRICULTURE, FISHING & TRAPPING	95	78	83	86	91	91	92	93
2 FORESTRY	13	12	12	12	12	12	12	13
3 MINERAL FUEL MINES & WELLS	11	12	12	12	30	55	56	57
4 OTHER MINES & QUARRIES	35	25	27	28	31	31	31	31
5 FOOD, FEED, BEVERAGES & TOBACCO	109	104	108	110	112	113	114	117
6 TEXTILES & CLOTHING	22	21	21	22	23	24	24	25
7 WOOD & FURNITURE	17	13	14	14	17	17	17	18
8 PAPER, PRINTING & ALLIED INDUSTRIES	79	61	67	71	78	80	82	84
9 PRIMARY METAL & METAL FABRICATING	28	26	26	26	28	28	28	29
10 MOTOR VEHICLES & PARTS	5	5	5	6	6	6	6	6
11 MACHINERY & OTH. TRANSPORTATION EQUIP.	31	27	27	27	28	29	29	31
12 ELECTRICAL PRODUCTS	14	6	8	10	12	13	13	14
13 CHEMICAL, RUBBER & PETROLEUM PRODUCTS	73	62	71	82	93	101	107	116
14 NON-METALLIC MINERAL PRODUCTS	11	10	10	11	12	12	11	12
15 OTHER MANUFACTURING INDUSTRIES	3	2	3	3	3	3	3	3
TOTAL MANUFACTURING	392	337	358	381	411	425	435	453
16 CONSTRUCTION	191	174	187	211	218	218	205	211
17 ELECTRIC POWER, WATER & GAS UTILITIES	83	81	85	90	95	99	103	108
TOTAL GOODS-PRODUCING INDUSTRIES	820	719	764	820	888	931	935	965
18 TRANSPORTATION & STORAGE	185	172	177	185	194	202	205	210
19 COMMUNICATION	133	131	140	151	163	176	188	201
20 TRADE	362	248	255	266	281	296	308	322
21 FINANCE, INSURANCE & REAL ESTATE	370	360	373	392	414	435	454	473
22 OTHER SERVICE INDUSTRIES	351	342	353	373	394	416	435	455
TOTAL SERVICE INDUSTRIES	1402	1252	1298	1367	1447	1525	1590	1661
23 GOVERNMENT SECTOR	690	700	710	721	731	744	757	771
TOTAL: ALL SECTORS (GROSS PROV. PRODUCT AT FACTOR COST - REAL)	2912	2671	2772	2908	3065	3200	3281	3398

Appendix Table 16.1 (cont'd.)
INSTITUTE FOR POLICY ANALYSIS, UNIVERSITY OF TORONTO
PRISM82A *** REVISED LO-TREND BASE CASE *** SEPT/82

NOVA SCOTIA
REAL DOMESTIC PRODUCT BY INDUSTRY
(MILLIONS OF 1971 DOLLARS)

	1989	1990	1991	1992	1993	1994	1995
1 AGRICULTURE,FISHING & TRAPPING	95	97	97	97	98	98	98
2 FORESTRY	13	13	14	14	14	14	14
3 MINERAL FUEL MINES & WELLS	57	58	59	76	65	75	76
4 OTHER MINES & QUARRIES	31	33	33	33	33	33	33
5 FOOD,FEED,BEVERAGES & TOBACCO	119	121	122	123	124	125	126
6 TEXTILES & CLOTHING	25	26	26	25	25	25	25
7 WOOD & FURNITURE	18	19	19	20	20	20	21
8 PAPER,PRINTING & ALLIED INDUSTRIES	87	91	92	95	97	99	100
9 PRIMARY METAL & METAL FABRICATING	29	30	31	31	32	31	31
10 MOTOR VEHICLES & PARTS	6	6	7	7	7	7	7
11 MACHINERY & OTH. TRANSPORTATION EQUIP.	31	33	33	34	34	33	33
12 ELECTRICAL PRODUCTS	14	15	16	16	16	16	17
13 CHEMICAL,RUBBER & PETROLEUM PRODUCTS	125	136	145	153	162	168	176
14 NON-METALLIC MINERAL PRODUCTS	12	13	13	14	14	14	14
15 OTHER MANUFACTURING INDUSTRIES	3	3	3	3	3	3	3
TOTAL MANUFACTURING	469	493	507	519	534	541	553
16 CONSTRUCTION	219	241	251	258	256	260	270
17 ELECTRIC POWER,WATER & GAS UTILITIES	113	119	124	128	134	138	144
TOTAL GOODS-PRODUCING INDUSTRIES	997	1054	1089	1125	1134	1159	1188
18 TRANSPORTATION & STORAGE	215	225	229	234	240	243	247
19 COMMUNICATION	215	233	247	262	278	293	314
20 TRADE	334	352	363	373	387	399	414
21 FINANCE,INSURANCE & REAL ESTATE	492	516	532	547	563	574	599
22 OTHER SERVICE INDUSTRIES	476	503	523	542	564	580	616
TOTAL SERVICE INDUSTRIES	1733	1829	1895	1958	2032	2088	2190
23 GOVERNMENT SECTOR	786	800	815	830	845	860	876
TOTAL: ALL SECTORS (GROSS PROV. PRODUCT AT FACTOR COST - REAL)	3516	3683	3799	3913	4010	4108	4254

INSTITUTE FOR POLICY ANALYSIS, UNIVERSITY OF TORONTO
PRISM82A *** REVISED LO-TREND BASE CASE *** SEPT/82

Appendix Table 16.1 (cont'd.)
REAL DOMESTIC PRODUCT BY INDUSTRY
NOVA SCOTIA

(YEAR-OVER-YEAR PER-CENT CHANGE)

	1981	1982	1983	1984	1985	1986	1987	1988
1 AGRICULTURE,FISHING & TRAPPING	24.32	-17.84	6.33	4.30	5.32	0.63	0.64	1.59
2 FORESTRY	-7.58	-1.88	-0.50	-3.66	2.45	-0.59	1.05	2.26
3 MINERAL FUEL MINES & WELLS	-10.11	3.17	3.22	0.20	151.32	80.29	2.50	1.18
4 OTHER MINES & QUARRIES	-24.01	-28.35	7.02	5.14	9.44	0.14	-0.54	1.19
5 FOOD,FEED,BEVERAGES & TOBACCO	4.37	-4.60	2.99	2.25	1.87	1.13	0.86	2.06
6 TEXTILES & CLOTHING	0.55	-3.35	0.10	4.90	4.92	3.40	0.53	2.46
7 WOOD & FURNITURE	-7.42	-26.36	7.20	6.44	16.43	1.68	0.78	2.12
8 PAPER,PRINTING & ALLIED INDUSTRIES	1.21	-22.14	8.64	6.51	9.41	2.86	2.35	3.13
9 PRIMARY METAL & METAL FABRICATING	2.14	-9.32	0.97	1.87	4.75	1.51	-0.24	2.19
10 MOTOR VEHICLES & PARTS	3.71	-7.61	-1.47	20.10	7.13	2.58	-2.11	0.42
11 MACHINERY & OTH. TRANSPORTATION EQUIP.	8.01	-13.08	0.83	-0.29	2.87	4.62	2.04	3.72
12 ELECTRICAL PRODUCTS	0.16	-56.92	31.74	25.79	23.69	5.74	3.17	4.74
13 CHEMICAL,RUBBER & PETROLEUM PRODUCTS	22.07	-15.18	14.77	14.72	14.35	8.01	6.31	8.36
14 NON-METALLIC MINERAL PRODUCTS	4.58	-10.77	1.21	4.91	8.18	0.69	-1.71	2.78
15 OTHER MANUFACTURING INDUSTRIES	9.10	-10.38	2.38	2.23	3.24	2.47	0.24	2.15
TOTAL MANUFACTURING	5.74	-14.08	6.23	6.45	7.77	3.59	2.37	4.03
16 CONSTRUCTION	13.76	-9.01	7.40	12.83	3.23	-0.12	-5.70	2.56
17 ELECTRIC POWER,WATER & GAS UTILITIES	0.53	-3.04	5.33	5.61	5.71	4.55	3.58	4.57
TOTAL GOODS-PRODUCING INDUSTRIES	6.50	-12.38	6.28	7.38	8.22	4.93	0.34	3.24
18 TRANSPORTATION & STORAGE	2.42	-7.19	3.10	4.81	4.53	4.04	1.77	2.29
19 COMMUNICATION	8.13	-1.83	6.61	7.96	8.53	7.83	6.69	7.12
20 TRADE	2.62	-31.48	2.93	4.39	5.55	5.23	4.11	4.48
21 FINANCE,INSURANCE & REAL ESTATE	4.57	-2.70	3.58	5.03	5.74	5.06	4.25	4.28
22 OTHER SERVICE INDUSTRIES	7.97	-2.74	3.38	5.54	5.60	5.61	4.49	4.72
TOTAL SERVICE INDUSTRIES	4.92	-10.65	3.65	5.32	5.81	5.42	4.24	4.52
23 GOVERNMENT SECTOR	1.66	1.48	1.40	1.54	1.35	1.81	1.80	1.87
TOTAL: ALL SECTORS (GROSS PROV. PRODUCT AT FACTOR COST - REAL)	4.56	-8.26	3.77	4.92	5.38	4.42	2.54	3.54

INSTITUTE FOR POLICY ANALYSIS, UNIVERSITY OF TORONTO
PRISM82A *** REVISED LO-TREND BASE CASE *** SEPT/82

Appendix Table 16.1 (cont'd.)
NOVA SCOTIA
REAL DOMESTIC PRODUCT BY INDUSTRY
(YEAR-OVER-YEAR PER-CENT CHANGE)

	1989	1990	1991	1992	1993	1994	1995
1 AGRICULTURE,FISHING & TRAPPING	1.45	2.16	-0.08	0.37	1.26	-0.31	0.03
2 FORESTRY	2.56	4.62	1.80	2.09	0.84	0.97
3 MINERAL FUEL MINES & WELLS	0.0	4.73	13.12	16.77	-15.19	16.91	1.04
4 OTHER MINES & QUARRIES	1.54	3.67	0.26	0.89	0.81	-0.64	-0.86
5 FOOD,FEED,BEVERAGES & TOBACCO	1.73	2.36	0.48	0.42	1.53	0.09	1.12
6 TEXTILES & CLOTHING	1.59	2.26	0.23	-0.80	-0.53	-1.55	-0.45
7 WOOD & FURNITURE	2.56	1.18	1.57	1.95	1.26	1.18	1.35
8 PAPER,PRINTING & ALLIED INDUSTRIES	3.26	4.32	1.77	2.43	2.70	1.43	1.30
9 PRIMARY METAL & METAL FABRICATING	1.97	1.87	1.60	1.23	1.03	-0.68	-0.15
10 MOTOR VEHICLES & PARTS	-0.07	3.95	5.11	0.15	1.29	0.07	-1.76
11 MACHINERY & OTH. TRANSPORTATION EQUIP.	1.59	4.68	2.28	1.33	0.19	-1.55	-0.39
12 ELECTRICAL PRODUCTS	2.95	3.86	2.32	1.98	1.21	0.69	2.56
13 CHEMICAL,RUBBER & PETROLEUM PRODUCTS	7.52	9.03	6.21	5.61	5.92	3.76	5.10
14 NON-METALLIC MINERAL PRODUCTS	1.30	1.72	2.85	1.61	1.39	0.16	1.91
15 OTHER MANUFACTURING INDUSTRIES	0.07	1.00	-1.40	-2.16	-2.24	-4.16	-1.38
TOTAL MANUFACTURING	3.57	5.17	2.70	2.44	2.89	1.25	2.17
16 CONSTRUCTION	4.17	9.82	4.05	2.96	-0.88	1.62	3.97
17 ELECTRIC POWER,WATER & GAS UTILITIES	4.51	5.75	3.74	3.50	4.57	3.21	4.40
TOTAL GOODS-PRODUCING INDUSTRIES	3.31	5.72	3.35	3.30	0.78	2.25	2.49
18 TRANSPORTATION & STORAGE	2.56	4.40	2.03	1.99	2.55	1.25	1.82
19 COMMUNICATION	6.82	8.14	6.24	5.91	6.25	5.22	7.11
20 TRADE	3.93	5.15	3.22	2.82	3.68	3.02	3.99
21 FINANCE, INSURANCE & REAL ESTATE	3.96	4.56	3.03	2.77	2.92	3.96	4.33
22 OTHER SERVICE INDUSTRIES	4.57	5.78	3.92	3.65	4.02	2.92	6.09
TOTAL SERVICE INDUSTRIES	4.29	5.55	3.60	3.34	3.77	2.79	4.85
23 GOVERNMENT SECTOR	1.90	1.82	1.81	1.86	1.78	1.81	1.83
TOTAL: ALL SECTORS (GROSS PROV. PRODUCT AT FACTOR COST - REAL)	3.47	4.76	3.14	3.01	2.49	2.43	3.55

INSTITUTE FOR POLICY ANALYSIS, UNIVERSITY OF TORONTO
PRISM82A *** REVISED LO-TREND BASE CASE *** SEPT/82

Appendix Table 16.2
REAL DOMESTIC PRODUCT BY INDUSTRY
QUEBEC

(MILLIONS OF 1971 DOLLARS)

	1981	1982	1983	1984	1985	1986	1987	1988
1 AGRICULTURE, FISHING & TRAPPING	426	407	412	410	413	412	412	415
2 FORESTRY	141	136	134	127	128	126	125	127
3 MINERAL FUEL MINES & WELLS	0	0	0	0	0	0	0	0
4 OTHER MINES & QUARRIES	269	205	217	209	211	205	199	196
5 FOOD, FEED, BEVERAGES & TOBACCO	997	946	969	985	998	1003	1006	1022
6 TEXTILES & CLOTHING	958	849	878	915	953	978	977	994
7 WOOD & FURNITURE	386	350	354	358	377	383	386	395
8 PAPER, PRINTING & ALLIED INDUSTRIES	1118	1056	1071	1070	1100	1117	1129	1149
9 PRIMARY METAL & METAL FABRICATING	1108	931	986	1011	1065	1088	1092	1122
10 MOTOR VEHICLES & PARTS	118	54	73	112	122	128	127	129
11 MACHINERY & OTH. TRANSPORTATION EQUIP.	725	629	634	631	648	677	690	714
12 ELECTRICAL PRODUCTS	426	353	348	347	353	356	351	351
13 CHEMICAL, RUBBER & PETROLEUM PRODUCTS	748	710	732	761	795	815	825	852
14 NON-METALLIC MINERAL PRODUCTS	226	199	200	207	222	221	215	219
15 OTHER MANUFACTURING INDUSTRIES	255	226	229	232	237	240	238	241
TOTAL MANUFACTURING	7064	6303	6475	6628	6870	7006	7035	7187
16 CONSTRUCTION	1397	831	942	1104	1186	1152	1092	1088
17 ELECTRIC POWER, WATER & GAS UTILITIES	1171	1130	1184	1244	1308	1361	1402	1459
TOTAL GOODS-PRODUCING INDUSTRIES	10468	9012	9364	9723	10116	10263	10266	10472
18 TRANSPORTATION & STORAGE	1656	1524	1560	1621	1681	1733	1751	1777
19 COMMUNICATION	1098	1078	1148	1237	1340	1442	1535	1642
20 TRADE	3745	3410	3482	3595	3749	3882	3981	4100
21 FINANCE, INSURANCE & REAL ESTATE	3496	2920	2996	3111	3249	3374	3476	3584
22 OTHER SERVICE INDUSTRIES	4879	4733	4863	5092	5331	5569	5758	5970
TOTAL SERVICE INDUSTRIES	14874	13665	14048	14657	15349	16000	16501	17073
23 GOVERNMENT SECTOR	3041	3064	3076	3087	3088	3117	3143	3174
TOTAL: ALL SECTORS (GROSS PROV. PRODUCT AT FACTOR COST - REAL)	28383	25742	26487	27467	28553	29379	29911	30719

INSTITUTE FOR POLICY ANALYSIS, UNIVERSITY OF TORONTO
PRISM82A *** REVISED LO-TREND BASE CASE *** SEPT/82

Appendix Table 16.2 (cont'd.)
QUEBEC
REAL DOMESTIC PRODUCT BY INDUSTRY
(MILLIONS OF 1971 DOLLARS)

	1989	1990	1991	1992	1993	1994	1995
1 AGRICULTURE,FISHING & TRAPPING	418	424	421	419	421	417	414
2 FORESTRY	128	132	132	133	134	133	132
3 MINERAL FUEL MINES & WELLS	0	0	0	0	0	0	0
4 OTHER MINES & QUARRIES	194	196	192	189	185	179	173
5 FOOD,FEED,BEVERAGES & TOBACCO	1034	1054	1053	1051	1062	1057	1063
6 TEXTILES & CLOTHING	1002	1018	1013	866	986	963	952
7 WOOD & FURNITURE	405	426	432	444	451	456	462
8 PAPER,PRINTING & ALLIED INDUSTRIES	1171	1206	1210	1223	1239	1240	1236
9 PRIMARY METAL & METAL FABRICATING	1151	1214	1240	1263	1283	1281	1286
10 MOTOR VEHICLES & PARTS	131	138	148	150	154	156	155
11 MACHINERY & OTH. TRANSPORTATION EQUIP.	724	751	773	781	787	773	758
12 ELECTRICAL PRODUCTS	345	350	346	338	331	320	315
13 CHEMICAL,RUBBER & PETROLEUM PRODUCTS	875	913	930	941	957	955	965
14 NON-METALLIC MINERAL PRODUCTS	223	238	242	244	245	242	244
15 OTHER MANUFACTURING INDUSTRIES	238	238	232	225	217	206	201
TOTAL MANUFACTURING	7301	7550	7618	7654	7712	7649	7651
16 CONSTRUCTION	1100	1171	1181	1177	1168	1147	1151
17 ELECTRIC POWER,WATER & GAS UTILITIES	1517	1596	1646	1695	1763	1810	1879
TOTAL GOODS-PRODUCING INDUSTRIES	10658	11070	11190	11266	11382	11335	11401
18 TRANSPORTATION & STORAGE	1808	1873	1895	1917	1951	1959	1978
19 COMMUNICATION	1751	1891	2006	2121	2250	2363	2527
20 TRADE	4204	4371	4456	4520	4629	4703	4826
21 FINANCE,INSURANCE & REAL ESTATE	3687	3832	3908	3976	4046	4079	4209
22 OTHER SERVICE INDUSTRIES	6185	6489	6684	6863	7078	7214	7582
TOTAL SERVICE INDUSTRIES	17634	18455	18950	19393	19954	20318	21122
23 GOVERNMENT SECTOR	3208	3242	3274	3306	3336	3366	3397
TOTAL: ALL SECTORS (GROSS PROV. PRODUCT AT FACTOR COST - REAL)	31500	32767	33414	33965	34673	35019	35920

INSTITUTE FOR POLICY ANALYSIS, UNIVERSITY OF TORONTO
PRISM82A *** REVISED LO-TREND BASE CASE *** SEPT/82

Appendix Table 16.2 (cont'd.)
REAL DOMESTIC PRODUCT BY INDUSTRY
QUEBEC

(YEAR-OVER-YEAR PER-CENT CHANGE)

	1981	1982	1983	1984	1985	1986	1987	1988
1 AGRICULTURE,FISHING & TRAPPING	-6.45	-4.46	1.30	-0.49	0.61	-0.11	-0.11	0.84
2 FORESTRY	-0.69	-3.20	-1.84	-4.96	1.07	-1.93	-0.32	0.88
3 MINERAL FUEL MINES & WELLS	0.0	0.0	0.0	0.0	0.0	0.0	0.0	0.0
4 OTHER MINES & QUARRIES	-7.39	-23.72	5.71	-3.55	0.75	-2.40	-3.06	-1.39
5 FOOD,FEED,BEVERAGES & TOBACCO	4.26	-5.19	2.48	1.68	1.25	0.51	0.37	1.55
6 TEXTILES & CLOTHING	3.49	-11.42	3.46	4.17	4.18	2.67	-0.18	1.73
7 WOOD & FURNITURE	1.44	-9.31	1.32	0.93	5.49	1.68	0.78	2.11
8 PAPER,PRINTING & ALLIED INDUSTRIES	6.10	-5.56	1.50	-0.18	2.83	1.56	1.05	1.81
9 PRIMARY METAL & METAL FABRICATING	-4.66	-15.90	5.89	2.49	5.38	2.12	0.35	2.79
10 MOTOR VEHICLES & PARTS	6.94	-54.55	36.45	53.43	8.92	4.25	-0.57	1.97
11 MACHINERY & OTH. TRANSPORTATION EQUIP.	-1.29	-13.20	0.69	-0.44	2.71	4.45	1.87	3.54
12 ELECTRICAL PRODUCTS	11.42	-17.15	-1.39	-0.24	1.88	0.82	-1.54	0.04
13 CHEMICAL,RUBBER & PETROLEUM PRODUCTS	7.38	-5.05	3.07	3.96	4.41	2.53	1.21	3.34
14 NON-METALLIC MINERAL PRODUCTS	14.86	-11.72	0.13	3.80	7.03	-0.38	-2.75	1.69
15 OTHER MANUFACTURING INDUSTRIES	11.77	-11.33	1.29	1.15	2.15	1.38	-0.83	1.07
TOTAL MANUFACTURING	3.49	-10.78	2.73	2.37	3.65	1.98	0.41	2.17
16 CONSTRUCTION	8.01	-40.48	13.34	17.20	7.36	-2.82	-5.19	-0.41
17 ELECTRIC POWER,WATER & GAS UTILITIES	2.56	-3.53	4.80	5.07	5.17	4.02	3.05	4.03
TOTAL GOODS-PRODUCING INDUSTRIES	3.14	-13.91	3.90	3.84	4.04	1.45	0.03	2.01
18 TRANSPORTATION & STORAGE	2.34	-7.98	2.32	3.96	3.65	3.15	1.00	1.50
19 COMMUNICATION	8.09	-1.86	6.47	7.77	8.30	7.62	6.48	6.92
20 TRADE	2.42	-8.94	2.11	3.26	4.27	3.56	2.53	3.00
21 FINANCE,INSURANCE & REAL ESTATE	4.43	-16.47	2.59	3.85	4.42	3.84	3.04	3.11
22 OTHER SERVICE INDUSTRIES	7.79	-2.99	2.74	4.71	4.69	4.46	3.40	3.69
TOTAL SERVICE INDUSTRIES	5.01	-8.13	2.80	4.34	4.72	4.24	3.13	3.47
23 GOVERNMENT SECTOR	1.61	0.75	0.37	0.38	0.02	0.93	0.86	0.96
TOTAL: ALL SECTORS (GROSS PROV. PRODUCT AT FACTOR COST - REAL)	3.94	-9.31	2.90	3.70	3.95	2.90	1.81	2.70

INSTITUTE FOR POLICY ANALYSIS, UNIVERSITY OF TORONTO
PRISM82A * REVISED LO-TREND BASE CASE *** SEPT/82**

Appendix Table 16.2 (cont'd.)
REAL DOMESTIC PRODUCT BY INDUSTRY
QUEBEC

(YEAR-OVER-YEAR PER-CENT CHANGE)

	1989	1990	1991	1992	1993	1994	1995
1 AGRICULTURE,FISHING & TRAPPING	0.70	1.39	-0.82	-0.37	0.51	-1.05	-0.71
2 FORESTRY	1.18	3.21	-0.11	0.43	0.71	-0.53	-0.39
3 MINERAL FUEL MINES & WELLS	0.0	0.0	0.0	0.0	0.0	0.0	0.0
4 OTHER MINES & QUARRIES	-1.05	1.02	-2.31	-1.70	-1.79	-3.21	-3.44
5 FOOD,FEED,BEVERAGES & TOBACCO	1.21	1.86	-0.08	-0.18	1.10	-0.52	0.58
6 TEXTILES & CLOTHING	0.87	1.53	-0.48	-1.50	-1.23	-2.24	-1.15
7 WOOD & FURNITURE	2.57	5.17	1.57	1.95	2.26	1.18	1.35
8 PAPER,PRINTING & ALLIED INDUSTRIES	1.93	2.96	0.37	1.06	1.32	0.06	-0.10
9 PRIMARY METAL & METAL FABRICATING	2.56	5.48	2.18	1.81	1.60	-0.13	0.40
10 MOTOR VEHICLES & PARTS	1.42	5.47	6.60	1.53	2.66	1.37	-0.50
11 MACHINERY & OTH. TRANSPORTATION EQUIP.	1.41	4.49	2.09	1.13	0.70	-1.77	-0.61
12 ELECTRICAL PRODUCTS	-1.68	1.39	-1.18	-2.18	-2.00	-3.51	-1.44
13 CHEMICAL,RUBBER & PETROLEUM PRODUCTS	2.71	4.29	1.84	1.19	1.71	-0.15	1.04
14 NON-METALLIC MINERAL PRODUCTS	2.20	6.57	1.76	0.83	0.31	-1.22	0.83
15 OTHER MANUFACTURING INDUSTRIES	-0.99	-0.07	-2.45	-3.20	-3.27	-5.18	-2.42
TOTAL MANUFACTURING	1.58	3.42	0.90	0.47	0.75	-0.81	0.03
16 CONSTRUCTION	1.08	6.49	0.83	-0.31	-0.81	-1.78	0.40
17 ELECTRIC POWER,WATER & GAS UTILITIES	3.97	5.19	3.19	2.95	4.01	2.66	3.83
TOTAL GOODS-PRODUCING INDUSTRIES	1.77	3.86	1.09	0.68	1.03	-0.42	0.58
18 TRANSPORTATION & STORAGE	1.75	3.60	1.19	1.13	1.80	0.38	1.00
19 COMMUNICATION	6.65	8.00	6.08	5.73	6.09	5.04	6.93
20 TRADE	2.54	3.96	1.95	1.43	2.42	1.60	2.60
21 FINANCE,INSURANCE & REAL ESTATE	2.86	3.94	1.98	1.65	1.86	0.81	3.18
22 OTHER SERVICE INDUSTRIES	3.59	4.91	3.02	2.67	3.12	1.93	5.10
TOTAL SERVICE INDUSTRIES	3.29	4.66	2.68	2.34	2.89	1.83	3.95
23 GOVERNMENT SECTOR	1.09	1.06	0.98	0.97	0.93	0.89	0.93
TOTAL: ALL SECTORS	2.54	4.02	1.97	1.65	2.08	1.00	2.57

(GROSS PROV. PRODUCT AT FACTOR COST - REAL)

INSTITUTE FOR POLICY ANALYSIS, UNIVERSITY OF TORONTO
PRISM82A *** REVISED LO-TREND BASE CASE *** SEPT/82

Appendix Table 16.3
REAL DOMESTIC PRODUCT BY INDUSTRY
ONTARIO

(MILLIONS OF 1971 DOLLARS)

	1981	1982	1983	1984	1985	1986	1987	1988
1 AGRICULTURE, FISHING & TRAPPING	969	933	952	955	968	974	980	996
2 FORESTRY	106	103	101	97	98	96	96	97
3 MINERAL FUEL MINES & WELLS	3	3	3	3	3	3	3	3
4 OTHER MINES & QUARRIES	485	449	432	416	419	410	399	395
5 FOOD, FEED, BEVERAGES & TOBACCO	1595	1535	1573	1603	1629	1641	1651	1680
6 TEXTILES & CLOTHING	688	670	675	714	754	785	795	820
7 WOOD & FURNITURE	402	361	363	363	379	382	382	387
8 PAPER, PRINTING & ALLIED INDUSTRIES	1417	1343	1368	1370	1414	1441	1462	1495
9 PRIMARY METAL & METAL FABRICATING	2540	2372	2345	2378	2480	2506	2488	2532
10 MOTOR VEHICLES & PARTS	1836	1750	1702	2018	2158	2210	2159	2165
11 MACHINERY & OTH. TRANSPORTATION EQUIP.	1321	1131	1124	1104	1119	1154	1160	1186
12 ELECTRICAL PRODUCTS	1248	1057	1065	1086	1130	1164	1171	1196
13 CHEMICAL, RUBBER & PETROLEUM PRODUCTS	1649	1574	1630	1702	1786	1840	1871	1943
14 NON-METALLIC MINERAL PRODUCTS	471	418	420	438	471	471	460	469
15 OTHER MANUFACTURING INDUSTRIES	677	607	621	635	656	672	673	688
TOTAL MANUFACTURING	13844	12816	12885	13411	13976	14266	14273	14560
16 CONSTRUCTION	1817	1962	1834	1722	1872	1844	1772	1790
17 ELECTRIC POWER, WATER & GAS UTILITIES	1304	1257	1316	1382	1452	1509	1553	1614
TOTAL GOODS-PRODUCING INDUSTRIES	18528	17524	17524	17985	18787	19102	19077	19456
18 TRANSPORTATION & STORAGE	1923	1794	1838	1916	1994	2063	2090	2128
19 COMMUNICATION	1718	1688	1800	1945	2114	2275	2424	2593
20 TRADE	5525	5198	5352	5601	5928	6155	6340	6561
21 FINANCE, INSURANCE & REAL ESTATE	6922	7208	7462	7843	8299	8652	8965	9296
22 OTHER SERVICE INDUSTRIES	6567	6395	6606	6979	7377	7721	8009	8332
TOTAL SERVICE INDUSTRIES	22656	22283	23058	24285	25712	26865	27828	28910
23 GOVERNMENT SECTOR	4934	5001	5063	5154	5237	5296	5365	5441
TOTAL: ALL SECTORS	46118	44807	45646	47424	49736	51264	52270	53807
(GROSS PROV. PRODUCT AT FACTOR COST - REAL)								

INSTITUTE FOR POLICY ANALYSIS, UNIVERSITY OF TORONTO
PRISM82A *** REVISED LO-TREND BASE CASE *** SEPT/82

Appendix Table 16.3 (cont'd.)

ONTARIO
REAL DOMESTIC PRODUCT BY INDUSTRY
(MILLIONS OF 1971 DOLLARS)

	1989	1990	1991	1992	1993	1994	1995
1 AGRICULTURE,FISHING & TRAPPING	1010	1032	1031	1035	1048	1045	1046
2 FORESTRY	99	102	102	103	104	104	104
3 MINERAL FUEL MINES & WELLS	3	3	3	3	4	4	4
4 OTHER MINES & QUARRIES	393	398	390	385	380	369	358
5 FOOD,FEED,BEVERAGES & TOBACCO	1704	1740	1743	1744	1747	1762	1776
6 TEXTILES & CLOTHING	839	863	871	870	871	863	865
7 WOOD & FURNITURE	393	410	417	417	423	424	426
8 PAPER,PRINTING & ALLIED INDUSTRIES	1530	1582	1595	1619	1648	1657	1664
9 PRIMARY METAL & METAL FABRICATING	2570	2683	2714	2735	2750	2719	2703
10 MOTOR VEHICLES & PARTS	2159	2240	2350	2350	2376	2373	2327
11 MACHINERY & OTH. TRANSPORTATION EQUIP.	1181	1224	1234	1232	1225	1188	1166
12 ELECTRICAL PRODUCTS	1201	1236	1256	1254	1255	1236	1244
13 CHEMICAL,RUBBER & PETROLEUM PRODUCTS	2005	2101	2150	2186	2235	2242	2276
14 NON-METALLIC MINERAL PRODUCTS	482	516	527	533	537	533	540
15 OTHER MANUFACTURING INDUSTRIES	889	695	685	671	656	628	620
TOTAL MANUFACTURING	14758	15299	15538	15611	15742	15626	15606
16 CONSTRUCTION	1836	1986	2034	2061	2080	2080	2127
17 ELECTRIC POWER,WATER & GAS UTILITIES	1677	1763	1818	1870	1944	1994	2069
TOTAL GOODS-PRODUCING INDUSTRIES	19777	20583	20917	21069	21302	21222	21314
18 TRANSPORTATION & STORAGE	2172	2257	2292	2325	2374	2390	2421
19 COMMUNICATION	2766	2988	3171	3354	3559	3739	3999
20 TRADE	6750	7042	7207	7337	7543	7692	7916
21 FINANCE,INSURANCE & REAL ESTATE	9066	10022	10201	10283	10757	10901	11303
22 OTHER SERVICE INDUSTRIES	8652	9100	9400	9676	10005	10223	10767
TOTAL SERVICE INDUSTRIES	29946	31419	32352	33197	34238	34945	36405
23 GOVERNMENT SECTOR	5517	5593	5667	5742	5815	5886	5957
TOTAL: ALL SECTORS (GROSS PROV. PRODUCT AT FACTOR COST - REAL)	55239	57594	58937	60008	61355	62053	63676

INSTITUTE FOR POLICY ANALYSIS, UNIVERSITY OF TORONTO
PRISM82A *** REVISED LO-TREND BASE CASE *** SEPT/82

Appendix Table 16.3 (cont'd.)
REAL DOMESTIC PRODUCT BY INDUSTRY
ONTARIO

(YEAR-OVER-YEAR PER-CENT CHANGE)

	1981	1982	1983	1984	1985	1986	1987	1988
1 AGRICULTURE, FISHING & TRAPPING	26.18	-3.74	2.06	0.26	1.37	0.64	0.65	1.60
2 FORESTRY	-9.90	-2.92	-1.56	-4.68	1.36	-1.65	-0.03	1.17
3 MINERAL FUEL MINES & WELLS	-7.99	3.17	3.22	0.20	0.53	3.02	2.50	1.18
4 OTHER MINES & QUARRIES	-8.15	-7.34	-3.81	-3.58	0.60	-2.05	-2.71	-1.02
5 FOOD, FEED, BEVERAGES & TOBACCO	4.28	-3.77	2.48	1.91	1.62	0.74	0.60	1.79
6 TEXTILES & CLOTHING	-1.60	-2.63	0.84	5.67	5.67	4.13	1.23	3.17
7 WOOD & FURNITURE	-7.52	-10.11	0.43	0.04	4.56	0.79	-0.11	1.22
8 PAPER, PRINTING & ALLIED INDUSTRIES	-0.51	-5.24	1.85	0.18	3.21	1.94	1.44	2.22
9 PRIMARY METAL & METAL FABRICATING	7.45	-6.63	-1.15	1.41	4.28	1.06	-0.69	1.73
10 MOTOR VEHICLES & PARTS	4.60	-4.67	-2.71	18.56	6.94	2.39	-2.29	0.24
11 MACHINERY & OTH. TRANSPORTATION EQUIP.	6.72	-14.35	-0.65	-1.75	1.36	3.08	0.54	2.20
12 ELECTRICAL PRODUCTS	7.10	-15.33	0.77	1.93	4.09	3.00	0.58	2.19
13 CHEMICAL, RUBBER & PETROLEUM PRODUCTS	1.80	-4.60	3.56	4.46	4.91	3.02	1.70	3.84
14 NON-METALLIC MINERAL PRODUCTS	0.57	-11.35	0.55	4.23	7.47	0.04	-2.35	2.11
15 OTHER MANUFACTURING INDUSTRIES	5.39	-10.38	2.38	2.23	3.24	2.47	0.24	2.15
TOTAL MANUFACTURING	3.77	-7.42	0.54	4.08	4.21	2.08	0.05	2.01
16 CONSTRUCTION	7.57	8.00	-6.52	-6.12	8.74	-1.53	-3.89	1.01
17 ELECTRIC POWER, WATER & GAS UTILITIES	7.17	-3.63	4.70	4.97	5.07	3.92	2.96	3.94
TOTAL GOODS-PRODUCING INDUSTRIES	4.90	-5.42	0.00	2.63	4.46	1.67	-0.13	1.99
18 TRANSPORTATION & STORAGE	2.36	-6.70	2.42	4.27	4.07	3.47	1.30	1.82
19 COMMUNICATION	7.79	-1.77	6.64	8.07	8.65	7.63	6.54	6.99
20 TRADE	1.26	-5.92	2.97	4.66	5.83	3.82	3.01	3.49
21 FINANCE, INSURANCE & REAL ESTATE	3.75	4.13	3.53	5.10	5.82	4.25	3.61	3.69
22 OTHER SERVICE INDUSTRIES	6.98	-2.62	3.30	5.64	5.70	4.66	3.74	4.03
TOTAL SERVICE INDUSTRIES	4.21	-1.64	3.48	5.32	5.88	4.49	3.58	3.89
23 GOVERNMENT SECTOR	0.42	1.35	1.25	1.80	1.60	1.14	1.31	1.42
TOTAL: ALL SECTORS (GROSS PROV. PRODUCT AT FACTOR COST - REAL)	4.06	-2.84	1.87	3.89	4.88	3.07	1.96	2.94

INSTITUTE FOR POLICY ANALYSIS, UNIVERSITY OF TORONTO
PRISM82A *** REVISED LO-TREND BASE CASE *** SEPT/82
Appendix Table 16.3 (cont'd.)

REAL DOMESTIC PRODUCT BY INDUSTRY
ONTARIO
(YEAR-OVER-YEAR PER-CENT CHANGE)

	1989	1990	1991	1992	1993	1994	1995
1 AGRICULTURE,FISHING & TRAPPING	1.46	2.16	-0.07	0.38	1.27	-0.30	0.03
2 FORESTRY	1.47	3.50	0.18	0.72	1.00	-0.24	-0.11
3 MINERAL FUEL MINES & WELLS	0.0	0.0	1.73	1.51	5.91	4.95	6.36
4 OTHER MINES & QUARRIES	-0.68	1.41	-1.93	-1.31	-1.39	-2.81	-3.03
5 FOOD,FEED,BEVERAGES & TOBACCO	1.43	2.10	0.18	0.05	1.33	-0.30	0.79
6 TEXTILES & CLOTHING	2.28	2.95	0.90	-0.15	0.12	-0.91	0.19
7 WOOD & FURNITURE	4.67	4.25	0.68	1.06	1.37	0.29	4.16
8 PAPER,PRINTING & ALLIED INDUSTRIES	2.35	3.40	0.82	1.52	1.80	0.54	0.40
9 PRIMARY METAL & METAL FABRICATING	1.51	4.40	0.78	1.14	0.57	-0.13	-0.60
10 MOTOR VEHICLES & PARTS	-0.26	3.77	4.92	-0.04	1.11	-0.12	-1.93
11 MACHINERY & OTH. TRANSPORTATION EQUIP.	0.09	3.14	0.78	-0.15	-0.57	-3.01	-1.98
12 ELECTRICAL PRODUCTS	0.43	0.92	-0.11	0.47	0.65
13 CHEMICAL,RUBBER & PETROLEUM PRODUCTS	3.20	4.80	2.34	1.67	2.20	0.33	1.52
14 NON-METALLIC MINERAL PRODUCTS	2.63	7.01	2.18	1.55	1.80	2.81	1.25
15 OTHER MANUFACTURING INDUSTRIES	0.07	1.00	-1.40	-2.16	-2.24	-4.16	-1.38
TOTAL MANUFACTURING	1.36	3.66	1.56	0.47	0.84	-0.74	-0.13
16 CONSTRUCTION	2.58	8.13	2.44	1.34	0.90	-0.01	2.29
17 ELECTRIC POWER,WATER & GAS UTILITIES	3.88	5.11	3.11	2.88	3.94	2.59	3.77
TOTAL GOODS-PRODUCING INDUSTRIES	1.65	4.08	1.62	0.73	1.10	-0.38	0.43
18 TRANSPORTATION & STORAGE	2.05	3.93	1.53	1.43	2.11	0.69	1.29
19 COMMUNICATION	6.67	8.03	6.12	5.76	6.12	5.07	6.94
20 TRADE	2.88	4.32	2.34	1.81	2.81	1.97	2.91
21 FINANCE,INSURANCE & REAL ESTATE	4.34	4.44	2.50	2.17	2.39	1.34	3.68
22 OTHER SERVICE INDUSTRIES	3.84	5.18	3.30	2.94	3.40	2.19	5.32
TOTAL SERVICE INDUSTRIES	3.58	4.92	2.97	2.61	3.13	2.07	4.18
23 GOVERNMENT SECTOR	1.39	1.37	1.34	1.31	1.28	1.22	1.20
TOTAL: ALL SECTORS (GROSS PROV. PRODUCT AT FACTOR COST - REAL)	2.66	4.26	2.33	1.82	2.24	1.14	2.61

INSTITUTE FOR POLICY ANALYSIS, UNIVERSITY OF TORONTO
PRISM82A *** REVISED LO-TREND BASE CASE *** SEPT/82

Appendix Table 16.4
REAL DOMESTIC PRODUCT BY INDUSTRY
ALBERTA

(MILLIONS OF 1971 DOLLARS)

	1981	1982	1983	1984	1985	1986	1987	1988
1 AGRICULTURE, FISHING & TRAPPING	656	636	655	662	676	686	696	712
2 FORESTRY	14	14	14	13	13	13	13	14
3 MINERAL FUEL MINES & WELLS	1158	1195	1234	1191	1179	1191	1177	1146
4 OTHER MINES & QUARRIES	267	259	262	265	280	288	293	304
5 FOOD, FEED, BEVERAGES & TOBACCO	226	218	226	231	236	239	242	248
6 TEXTILES & CLOTHING	27	26	26	28	29	30	30	31
7 WOOD & FURNITURE	93	86	88	90	97	100	102	105
8 PAPER, PRINTING & ALLIED INDUSTRIES	146	143	152	157	168	178	187	197
9 PRIMARY METAL & METAL FABRICATING	217	203	211	221	238	248	254	267
10 MOTOR VEHICLES & PARTS	24	22	22	27	28	29	29	29
11 MACHINERY & OTH. TRANSPORTATION EQUIP.	116	106	112	117	126	137	145	156
12 ELECTRICAL PRODUCTS	22	19	20	20	22	23	24	25
13 CHEMICAL, RUBBER & PETROLEUM PRODUCTS	228	221	232	246	261	273	282	297
14 NON-METALLIC MINERAL PRODUCTS	127	116	121	130	145	149	150	158
15 OTHER MANUFACTURING INDUSTRIES	20	19	19	20	21	22	23	24
TOTAL MANUFACTURING	1246	1179	1229	1287	1371	1429	1468	1537
16 CONSTRUCTION	1897	1613	1594	1652	1858	1897	1890	1935
17 ELECTRIC POWER, WATER & GAS UTILITIES	326	320	340	362	386	408	426	450
TOTAL GOODS-PRODUCING INDUSTRIES	5564	5217	5326	5433	5765	5912	5963	6098
18 TRANSPORTATION & STORAGE	1143	1069	1104	1159	1214	1264	1289	1321
19 COMMUNICATION	904	892	954	1031	1121	1208	1288	1379
20 TRADE	1727	1630	1711	1800	1917	2007	2082	2170
21 FINANCE, INSURANCE & REAL ESTATE	1579	1574	1654	1748	1860	1953	2034	2122
22 OTHER SERVICE INDUSTRIES	1918	1909	2000	2122	2254	2374	2476	2591
TOTAL SERVICE INDUSTRIES	7271	7074	7423	7859	8365	8806	9170	9584
23 GOVERNMENT SECTOR	1440	1508	1559	1596	1632	1666	1700	1738
TOTAL: ALL SECTORS (GROSS PROV. PRODUCT AT FACTOR COST - REAL)	14275	13798	14308	14887	15761	16384	16833	17421

INSTITUTE FOR POLICY ANALYSIS, UNIVERSITY OF TORONTO
PRISM82A *** REVISED LO-TREND BASE CASE *** SEPT/82

Appendix Table 16.4 (cont'd.)
ALBERTA
REAL DOMESTIC PRODUCT BY INDUSTRY
(MILLIONS OF 1971 DOLLARS)

	1989	1990	1991	1992	1993	1994	1995
1 AGRICULTURE,FISHING & TRAPPING	728	750	755	764	779	782	788
2 FORESTRY	14	15	15	15	16	16	16
3 MINERAL FUEL MINES & WELLS	1124	1143	1143	1192	1265	1305	1353
4 OTHER MINES & QUARRIES	317	336	344	354	364	368	372
5 FOOD,FEED,BEVERAGES & TOBACCO	253	260	263	265	271	273	277
6 TEXTILES & CLOTHING	31	32	32	32	32	31	31
7 WOOD & FURNITURE	110	117	120	124	128	132	135
8 PAPER,PRINTING & ALLIED INDUSTRIES	209	223	232	243	255	265	274
9 PRIMARY METAL & METAL FABRICATING	279	299	311	323	333	338	345
10 MOTOR VEHICLES & PARTS	29	30	31	31	32	32	31
11 MACHINERY & OTH. TRANSPORTATION EQUIP.	165	179	189	198	206	209	214
12 ELECTRICAL PRODUCTS	25	27	27	28	28	28	29
13 CHEMICAL,RUBBER & PETROLEUM PRODUCTS	311	331	343	354	367	373	384
14 NON-METALLIC MINERAL PRODUCTS	191	184	194	202	209	213	221
15 OTHER MANUFACTURING INDUSTRIES	24	25	25	25	25	24	24
TOTAL MANUFACTURING	1602	1706	1768	1824	1886	1917	1966
16 CONSTRUCTION	2035	2272	2369	2461	2580	2644	2771
17 ELECTRIC POWER,WATER & GAS UTILITIES	474	506	529	553	583	607	639
TOTAL GOODS-PRODUCING INDUSTRIES	6294	6728	6923	7162	7472	7639	7905
18 TRANSPORTATION & STORAGE	1358	1422	1453	1485	1529	1551	1583
19 COMMUNICATION	1473	1594	1693	1794	1908	2008	2153
20 TRADE	2250	2368	2445	2517	2621	2707	2826
21 FINANCE, INSURANCE & REAL ESTATE	2706	2520	2396	2469	2553	2613	2739
22 OTHER SERVICE INDUSTRIES	2706	2865	2981	3094	3229	3332	3457
TOTAL SERVICE INDUSTRIES	9994	10569	10968	11360	11840	12211	12848
23 GOVERNMENT SECTOR	1777	1818	1861	1908	1958	2010	2065
TOTAL: ALL SECTORS (GROSS PROV. PRODUCT AT FACTOR COST - REAL)	18064	19114	19752	20429	21270	21860	22818

INSTITUTE FOR POLICY ANALYSIS, UNIVERSITY OF TORONTO
PRISM82A   *** REVISED LO-TREND BASE CASE *** SEPT/82

Appendix Table 16.4 (cont'd.)
REAL DOMESTIC PRODUCT BY INDUSTRY
ALBERTA

(YEAR-OVER-YEAR PER-CENT CHANGE)

	1981	1982	1983	1984	1985	1986	1987	1988
1 AGRICULTURE, FISHING & TRAPPING	7.99	-2.94	2.90	1.07	2.19	1.44	1.44	2.39
2 FORESTRY	-1.10	-1.88	-0.50	-3.66	2.45	-0.59	1.05	2.26
3 MINERAL FUEL MINES & WELLS	-7.99	3.17	3.22	-3.43	-1.00	1.03	-1.21	-2.61
4 OTHER MINES & QUARRIES	-1.86	-2.81	0.96	1.23	5.60	2.78	2.03	3.71
5 FOOD, FEED, BEVERAGES & TOBACCO	4.50	-3.19	3.38	2.24	2.32	1.36	1.16	2.28
6 TEXTILES & CLOTHING	-6.61	-3.35	0.10	4.90	4.92	3.40	0.53	2.46
7 WOOD & FURNITURE	-4.92	-7.95	2.81	2.39	6.99	3.11	2.17	3.51
8 PAPER, PRINTING & ALLIED INDUSTRIES	7.05	-1.53	5.75	3.92	6.98	5.59	4.99	5.73
9 PRIMARY METAL & METAL FABRICATING	-1.99	-6.51	4.00	4.83	7.71	4.29	2.41	4.83
10 MOTOR VEHICLES & PARTS	5.39	-7.61	-1.47	20.10	7.13	2.58	-2.11	0.42
11 MACHINERY & OTH. TRANSPORTATION EQUIP.	-10.31	-8.69	5.66	4.26	7.35	8.97	6.11	7.68
12 ELECTRICAL PRODUCTS	19.68	-13.22	3.19	4.31	6.45	5.26	2.73	4.31
13 CHEMICAL, RUBBER & PETROLEUM PRODUCTS	-5.39	-3.21	5.07	5.99	6.45	4.53	3.19	5.35
14 NON-METALLIC MINERAL PRODUCTS	2.56	-8.19	4.03	7.75	11.02	3.26	0.72	5.25
15 OTHER MANUFACTURING INDUSTRIES	14.63	-8.50	4.49	4.29	5.28	4.45	2.14	4.05
TOTAL MANUFACTURING	-0.61	-5.31	4.17	4.78	6.53	4.19	2.70	4.70
16 CONSTRUCTION	10.10	-14.96	-1.23	3.67	12.45	2.09	-0.37	2.40
17 ELECTRIC POWER, WATER & GAS UTILITIES	-1.96	-2.11	6.33	6.60	6.69	5.52	4.53	5.52
TOTAL GOODS-PRODUCING INDUSTRIES	1.88	-6.24	2.09	2.01	6.11	2.56	0.87	2.26
18 TRANSPORTATION & STORAGE	2.41	-6.42	3.27	4.90	4.77	4.17	1.98	2.47
19 COMMUNICATION	8.51	-1.32	6.94	8.11	8.70	7.77	6.64	7.10
20 TRADE	5.31	-5.63	4.98	5.20	6.47	4.73	3.71	4.26
21 FINANCE, INSURANCE & REAL ESTATE	6.70	-0.32	5.11	5.65	6.44	4.97	4.19	4.33
22 OTHER SERVICE INDUSTRIES	9.91	-0.50	4.78	6.09	6.22	5.36	4.29	4.63
TOTAL SERVICE INDUSTRIES	6.70	-2.71	4.94	5.87	6.44	5.28	4.13	4.52
23 GOVERNMENT SECTOR	4.75	4.70	3.39	2.38	2.27	2.09	2.05	2.23
TOTAL: ALL SECTORS (GROSS PROV. PRODUCT AT FACTOR COST - REAL)	4.58	-3.34	3.69	4.05	5.87	3.95	2.74	3.49

INSTITUTE FOR POLICY ANALYSIS, UNIVERSITY OF TORONTO
PRISM82A *** REVISED LO-TREND BASE CASE *** SEPT/82

Appendix Table 16.4 (cont'd.)

REAL DOMESTIC PRODUCT BY INDUSTRY
ALBERTA

(YEAR-OVER-YEAR PER-CENT CHANGE)

	1989	1990	1991	1992	1993	1994	1995
1 AGRICULTURE,FISHING & TRAPPING	2.24	2.94	0.68	1.14	2.02	0.43	0.76
2 FORESTRY	2.56	4.62	1.26	1.80	2.09	0.84	0.97
3 MINERAL FUEL MINES & WELLS	-1.98	1.73	-0.03	4.28	6.13	3.16	3.74
4 OTHER MINES & QUARRIES	3.98	6.05	2.45	2.99	2.79	1.19	0.85
5 FOOD,FEED,BEVERAGES & TOBACCO	2.08	2.97	0.83	0.91	2.32	0.56	1.70
6 TEXTILES & CLOTHING	1.59	2.26	0.23	-0.80	-0.53	-1.55	-0.45
7 WOOD & FURNITURE	3.95	6.57	2.90	3.27	3.57	2.45	2.60
8 PAPER,PRINTING & ALLIED INDUSTRIES	5.80	6.81	4.08	4.76	4.98	3.63	3.44
9 PRIMARY METAL & METAL FABRICATING	4.54	7.44	4.02	3.59	3.32	1.52	2.01
10 MOTOR VEHICLES & PARTS	-0.07	3.95	5.11	0.15	1.29	0.07	-1.76
11 MACHINERY & OTH. TRANSPORTATION EQUIP.	5.32	8.38	5.76	4.66	4.11	1.46	2.57
12 ELECTRICAL PRODUCTS	2.45	5.59	2.86	1.76	1.89	0.28	2.38
13 CHEMICAL,RUBBER & PETROLEUM PRODUCTS	4.70	6.31	3.80	3.12	3.64	1.73	2.92
14 NON-METALLIC MINERAL PRODUCTS	5.71	10.16	5.13	4.11	3.52	1.89	3.95
15 OTHER MANUFACTURING INDUSTRIES	1.90	2.81	0.33	-0.47	-0.57	-2.56	0.24
TOTAL MANUFACTURING	4.28	6.49	3.60	3.18	3.41	1.66	2.54
16 CONSTRUCTION	5.16	11.65	4.26	3.88	4.85	2.47	4.82
17 ELECTRIC POWER,WATER & GAS UTILITIES	5.45	6.68	4.64	4.40	5.47	4.09	5.28
TOTAL GOODS-PRODUCING INDUSTRIES	3.21	6.88	2.90	3.46	4.33	2.23	3.49
18 TRANSPORTATION & STORAGE	2.76	4.70	2.22	2.21	2.91	1.46	2.08
19 COMMUNICATION	6.79	8.17	6.27	5.94	6.34	5.28	7.19
20 TRADE	3.66	5.23	3.28	2.94	4.11	3.28	4.38
21 FINANCE,INSURANCE & REAL ESTATE	3.96	5.17	3.25	3.06	3.40	2.35	4.82
22 OTHER SERVICE INDUSTRIES	4.45	5.88	4.02	3.80	4.38	3.17	6.44
TOTAL SERVICE INDUSTRIES	4.27	5.76	3.78	3.57	4.23	3.14	5.21
23 GOVERNMENT SECTOR	2.22	2.33	2.34	2.53	2.66	2.63	2.75
TOTAL: ALL SECTORS (GROSS PROV. PRODUCT AT FACTOR COST - REAL)	3.69	5.81	3.33	3.43	4.12	2.77	4.38

INSTITUTE FOR POLICY ANALYSIS, UNIVERSITY OF TORONTO
PRISM82A *** REVISED LO-TREND BASE CASE *** SEPT/82

Appendix Table 16.5
REAL DOMESTIC PRODUCT BY INDUSTRY
BRITISH COLUMBIA

(MILLIONS OF 1971 DOLLARS)

	1981	1982	1983	1984	1985	1986	1987	1988
1 AGRICULTURE, FISHING & TRAPPING	236	229	237	240	246	250	254	261
2 FORESTRY	403	353	368	387	416	434	442	455
3 MINERAL FUEL MINES & WELLS	129	133	137	186	187	192	197	200
4 OTHER MINES & QUARRIES	203	195	195	195	203	207	209	214
5 FOOD, FEED, BEVERAGES & TOBACCO	264	254	262	269	274	277	280	286
6 TEXTILES & CLOTHING	32	32	32	34	36	38	39	40
7 WOOD & FURNITURE	627	568	576	581	613	623	628	641
8 PAPER, PRINTING & ALLIED INDUSTRIES	542	525	547	560	590	615	637	665
9 PRIMARY METAL & METAL FABRICATING	307	278	281	286	300	304	304	310
10 MOTOR VEHICLES & PARTS	49	46	47	57	63	66	66	68
11 MACHINERY & OTH. TRANSPORTATION EQUIP.	178	158	163	166	175	187	194	206
12 ELECTRICAL PRODUCTS	31	26	26	27	28	29	29	29
13 CHEMICAL, RUBBER & PETROLEUM PRODUCTS	149	145	153	163	175	183	190	201
14 NON-METALLIC MINERAL PRODUCTS	80	72	74	78	86	88	87	91
15 OTHER MANUFACTURING INDUSTRIES	28	26	26	27	28	28	28	29
TOTAL MANUFACTURING	2287	2130	2187	2249	2367	2438	2482	2567
16 CONSTRUCTION	998	894	929	1083	1206	1195	1163	1204
17 ELECTRIC POWER, WATER & GAS UTILITIES	463	454	484	517	553	585	613	648
TOTAL GOODS-PRODUCING INDUSTRIES	4717	4389	4536	4857	5179	5302	5360	5548
18 TRANSPORTATION & STORAGE	1463	1362	1401	1494	1579	1638	1664	1699
19 COMMUNICATION	819	806	862	932	1013	1093	1166	1249
20 TRADE	1945	1802	1875	1970	2098	2201	2287	2384
21 FINANCE, INSURANCE & REAL ESTATE	1974	1938	2021	2131	2265	2379	2480	2584
22 OTHER SERVICE INDUSTRIES	2058	2021	2104	2230	2368	2498	2609	2728
TOTAL SERVICE INDUSTRIES	8259	7930	8263	8755	9323	9808	10206	10646
23 GOVERNMENT SECTOR	1660	1704	1744	1783	1823	1865	1907	1949
TOTAL: ALL SECTORS (GROSS PROV. PRODUCT AT FACTOR COST - REAL)	14636	14022	14544	15395	16325	16974	17473	18143

INSTITUTE FOR POLICY ANALYSIS, UNIVERSITY OF TORONTO
PRISM82A * REVISED LO-TREND BASE CASE *** SEPT/82**

Appendix Table 16.5 (cont'd.)
REAL DOMESTIC PRODUCT BY INDUSTRY
BRITISH COLUMBIA

(MILLIONS OF 1971 DOLLARS)

	1989	1990	1991	1992	1993	1994	1995
1 AGRICULTURE,FISHING & TRAPPING	268	276	279	283	289	291	294
2 FORESTRY	470	495	504	517	531	539	548
3 MINERAL FUEL MINES & WELLS	200	203	213	232	198	215	229
4 OTHER MINES & QUARRIES	220	232	235	241	246	247	248
5 FOOD,FEED,BEVERAGES & TOBACCO	291	299	300	302	307	307	311
6 TEXTILES & CLOTHING	41	43	43	43	44	43	44
7 WOOD & FURNITURE	658	692	703	717	733	741	751
8 PAPER,PRINTING & ALLIED INDUSTRIES	695	734	755	782	813	834	854
9 PRIMARY METAL & METAL FABRICATING	316	332	337	341	345	342	342
10 MOTOR VEHICLES & PARTS	70	74	80	82	84	86	86
11 MACHINERY & OTH. TRANSPORTATION EQUIP.	213	227	237	244	251	252	255
12 ELECTRICAL PRODUCTS	29	30	30	30	30	29	29
13 CHEMICAL,RUBBER & PETROLEUM PRODUCTS	211	225	235	243	253	258	267
14 NON-METALLIC MINERAL PRODUCTS	95	103	107	111	113	115	118
15 OTHER MANUFACTURING INDUSTRIES	29	29	29	28	28	26	26
TOTAL MANUFACTURING	2649	2788	2856	2923	3000	3034	3083
16 CONSTRUCTION	1265	1402	1472	1528	1580	1619	1697
17 ELECTRIC POWER,WATER & GAS UTILITIES	685	733	768	804	849	886	935
TOTAL GOODS-PRODUCING INDUSTRIES	5757	6129	6328	6528	6693	6832	7033
18 TRANSPORTATION & STORAGE	1740	1814	1847	1879	1925	1944	1976
19 COMMUNICATION	1334	1444	1534	1625	1728	1818	1948
20 TRADE	2471	2597	2678	2749	2850	2930	3042
21 FINANCE,INSURANCE & REAL ESTATE	2684	2817	2903	2982	3070	3128	3263
22 OTHER SERVICE INDUSTRIES	2848	3013	3130	3241	3371	3466	3675
TOTAL SERVICE INDUSTRIES	11077	11685	12092	12478	12943	13287	13904
23 GOVERNMENT SECTOR	1991	2035	2079	2124	2170	2216	2264
TOTAL: ALL SECTORS	18824	19848	20498	21130	21807	22335	23201

(GROSS PROV. PRODUCT AT FACTOR COST - REAL)

INSTITUTE FOR POLICY ANALYSIS, UNIVERSITY OF TORONTO
PRISM82A *** REVISED LO-TREND BASE CASE *** SEPT/82

Appendix Table 16.5 (cont'd.)
REAL DOMESTIC PRODUCT BY INDUSTRY
BRITISH COLUMBIA

(YEAR-OVER-YEAR PER-CENT CHANGE)

	1981	1982	1983	1984	1985	1986	1987	1988
1 AGRICULTURE, FISHING & TRAPPING	41.21	-2.66	3.19	1.36	2.47	1.72	1.71	2.66
2 FORESTRY	-1.80	-12.39	4.32	5.19	7.47	4.31	1.76	2.97
3 MINERAL FUEL MINES & WELLS	-7.97	3.17	3.22	35.47	0.53	3.02	2.50	1.18
4 OTHER MINES & QUARRIES	-2.67	-3.78	-0.14	0.07	4.39	1.61	0.90	2.63
5 FOOD, FEED, BEVERAGES & TOBACCO	4.16	-3.83	3.08	2.64	2.09	1.10	0.98	2.15
6 TEXTILES & CLOTHING	3.89	-2.07	1.40	6.25	6.24	4.69	1.77	3.70
7 WOOD & FURNITURE	1.81	-9.31	1.32	0.93	5.49	1.68	0.78	2.11
8 PAPER, PRINTING & ALLIED INDUSTRIES	-8.03	-3.12	4.12	2.38	5.46	4.15	3.62	4.40
9 PRIMARY METAL & METAL FABRICATING	-1.89	-9.32	0.97	1.87	4.75	1.51	-0.24	2.19
10 MOTOR VEHICLES & PARTS	14.90	-5.07	1.16	23.22	9.85	5.11	0.24	2.78
11 MACHINERY & OTH. TRANSPORTATION EQUIP.	17.24	-11.06	3.12	1.92	5.09	6.83	4.15	5.81
12 ELECTRICAL PRODUCTS	10.48	-15.75	0.28	1.45	3.60	2.52	0.12	1.73
13 CHEMICAL, RUBBER & PETROLEUM PRODUCTS	3.78	-2.61	5.68	6.56	6.99	5.03	3.65	5.80
14 NON-METALLIC MINERAL PRODUCTS	-9.44	-9.61	2.50	6.24	9.53	1.93	-0.51	4.01
15 OTHER MANUFACTURING INDUSTRIES	6.98	-10.38	2.38	2.23	3.24	2.47	0.24	2.15
TOTAL MANUFACTURING	0.18	-6.83	2.64	2.84	5.28	3.00	1.80	3.39
16 CONSTRUCTION	-2.36	-10.39	3.84	16.64	11.38	-0.93	-2.70	3.51
17 ELECTRIC POWER, WATER & GAS UTILITIES	10.50	-1.85	6.61	6.87	6.96	5.77	4.77	5.76
TOTAL GOODS-PRODUCING INDUSTRIES	1.47	-6.96	3.35	7.08	6.64	2.37	1.10	3.51
18 TRANSPORTATION & STORAGE	2.33	-6.89	2.87	6.63	5.69	3.76	1.60	2.11
19 COMMUNICATION	8.26	-1.54	6.85	8.12	8.75	7.85	6.72	7.13
20 TRADE	3.33	-7.34	4.06	5.04	6.48	4.92	3.94	4.24
21 FINANCE, INSURANCE & REAL ESTATE	5.07	-1.83	4.27	5.42	6.33	5.00	4.25	4.21
22 OTHER SERVICE INDUSTRIES	8.50	-1.80	4.11	5.95	6.21	5.48	4.44	4.59
TOTAL SERVICE INDUSTRIES	5.29	-3.99	4.20	5.95	6.48	5.20	4.06	4.30
23 GOVERNMENT SECTOR	2.57	2.61	2.39	2.19	2.27	2.27	2.27	2.19
TOTAL: ALL SECTORS (GROSS PROV. PRODUCT AT FACTOR COST - REAL)	3.72	-4.20	3.72	5.85	6.04	3.98	2.94	3.83

INSTITUTE FOR POLICY ANALYSIS, UNIVERSITY OF TORONTO
PRISM82A *** REVISED LO-TREND BASE CASE *** SEPT/82

Appendix Table 16.5 (cont'd.)

BRITISH COLUMBIA
REAL DOMESTIC PRODUCT BY INDUSTRY

(YEAR-OVER-YEAR PER-CENT CHANGE)

	1989	1990	1991	1992	1993	1994	1995
1 AGRICULTURE,FISHING & TRAPPING	2.51	3.20	0.94	1.39	2.27	0.68	1.00
2 FORESTRY	3.26	5.32	1.93	2.47	2.75	1.48	1.61
3 MINERAL FUEL MINES & WELLS	0.0	1.73	4.82	9.25	-14.96	8.77	6.36
4 OTHER MINES & QUARRIES	2.96	5.10	1.62	2.24	2.14	0.65	0.40
5 FOOD,FEED,BEVERAGES & TOBACCO	1.84	2.54	0.55	0.50	1.68	0.16	1.23
6 TEXTILES & CLOTHING	2.81	3.47	1.41	0.35	0.61	-0.43	0.66
7 WOOD & FURNITURE	2.57	1.17	2.57	1.95	2.26	1.18	1.35
8 PAPER,PRINTING & ALLIED INDUSTRIES	4.51	5.56	2.91	3.61	3.87	2.57	2.42
9 PRIMARY METAL & METAL FABRICATING	1.97	4.87	1.60	1.23	1.03	-0.68	-0.15
10 MOTOR VEHICLES & PARTS	2.22	7.58	7.41	2.69	2.42	2.12	2.22
11 MACHINERY & OTH. TRANSPORTATION EQUIP.	3.59	6.70	4.22	3.21	2.75	0.20	1.35
12 ELECTRICAL PRODUCTS	-0.02	2.10	0.49	-0.53	-1.35	-1.88	0.22
13 CHEMICAL,RUBBER & PETROLEUM PRODUCTS	5.13	6.72	4.19	3.49	4.01	2.08	3.27
14 NON-METALLIC MINERAL PRODUCTS	4.52	8.97	4.04	3.07	2.53	0.95	3.03
15 OTHER MANUFACTURING INDUSTRIES	0.07	1.00	-1.40	-2.16	-2.24	-4.16	-1.38
TOTAL MANUFACTURING	3.21	5.27	2.44	2.34	2.62	1.15	1.61
16 CONSTRUCTION	5.12	10.80	4.97	3.85	3.39	2.45	4.81
17 ELECTRIC POWER,WATER & GAS UTILITIES	5.68	6.92	4.87	4.62	5.69	4.30	5.48
TOTAL GOODS-PRODUCING INDUSTRIES	3.76	6.46	3.25	3.17	2.53	2.07	2.95
18 TRANSPORTATION & STORAGE	2.37	4.26	1.83	1.77	2.41	1.02	1.63
19 COMMUNICATION	6.82	8.19	6.28	6.94	6.30	5.25	7.14
20 TRADE	3.62	5.12	3.12	2.66	3.64	2.82	3.84
21 FINANCE,INSURANCE & REAL ESTATE	3.84	4.98	3.03	2.74	2.94	1.90	4.30
22 OTHER SERVICE INDUSTRIES	5.40	5.77	3.88	3.57	4.01	2.81	6.01
TOTAL SERVICE INDUSTRIES	4.05	5.49	3.48	3.19	3.73	2.66	4.64
23 GOVERNMENT SECTOR	2.16	2.19	2.16	2.20	2.15	2.12	2.16
TOTAL: ALL SECTORS (GROSS PROV. PRODUCT AT FACTOR COST - REAL)	3.76	5.44	3.28	3.08	3.20	2.42	3.88

List of tables

Chapter 2: The national outlook - FOCUS
1. National outlook assumptions 4
2. Provinces - major projects 5
3. National projection - summary Revised low-trend (per cent) 8
4. Components of real GNP - growth rates 10

Chapter 3: National industrial detail - PRISM
5. Growth rates by industry - Canada 14

Chapter 4: Provincial projection - PRISM
6. Real output shares by province 25
7. Labour productivity by province: all industries (per cent deviation from national) 28
8. Productivity - manufacturing (per cent deviation from national) 29
9. Productivity - goods producing (per cent deviation from national) 30
10. Growth of labour-force source population 33

Chapter 5: Provincial projection: overview of results
11. Growth rates of real output by province 37
12. Growth rates of GDP deflators by province 39
13. Shares of nominal GDP - by province 40
14. Unemployment rates - by province 41
15. Ratio of real disposable income per capita to national 43

16 Growth rates of real personal disposable income per capita - by province 45

Chapter 6: Results by province
17 Real GDP and GDP deflators by province 47
18 Employment and unemployment rates by province 51

Appendix - Tables 60

National

1 Summary of national projection
2 GNP in constant dollars
3 GNP deflators
4 Balance of payments
5 Labour market
6 Federal government
7 Combined public sector
8 National and personal income
9 Sources and disposition of saving

Energy

10 Oil and gas: production, demand, prices
11 Oil and gas: trade and price summary

Industrial Detail

12 RDP by industry
13 Employment by industry
14 Productivity by industry

Provincial Detail

15.1 - 15.10 Summary by province
16.1 - 16.5 Output by industry (selected provinces)

List of figures

Chapter 2: The national outlook - FOCUS
1   Canada: main indicators   9
2   Expenditure shares of GNP   11

Chapter 3: National industrial detail - PRISM
3   Shares of major industry groups in total output   15

Chapter 4: Provincial projection - PRISM
4   Output shares by province - other mining   19
5   Output shares by province - mineral fuels   20
6   Output shares by province - pulp and paper   21
7   Output shares by province - primary metal and metal fabricating   22
8   Output shares by province - construction   23
9   Output shares by province - chemicals and petroleum   24
10 Output shares by province - manufacturing   26
11 Output shares by province - goods producing   27
12 Productivity - ratio to national average   32
13 Population by region   34

Chapter 5: Provincial projection: overview of results
14 Share of total output - by regions   36
15 Output growth - by region   38
16 Unemployment rates - by region   42
17 Real personal disposable income per capita - ratio to national average   44